YAGI-ANTENNA DESIGN

by Dr James L. Lawson, W2PV

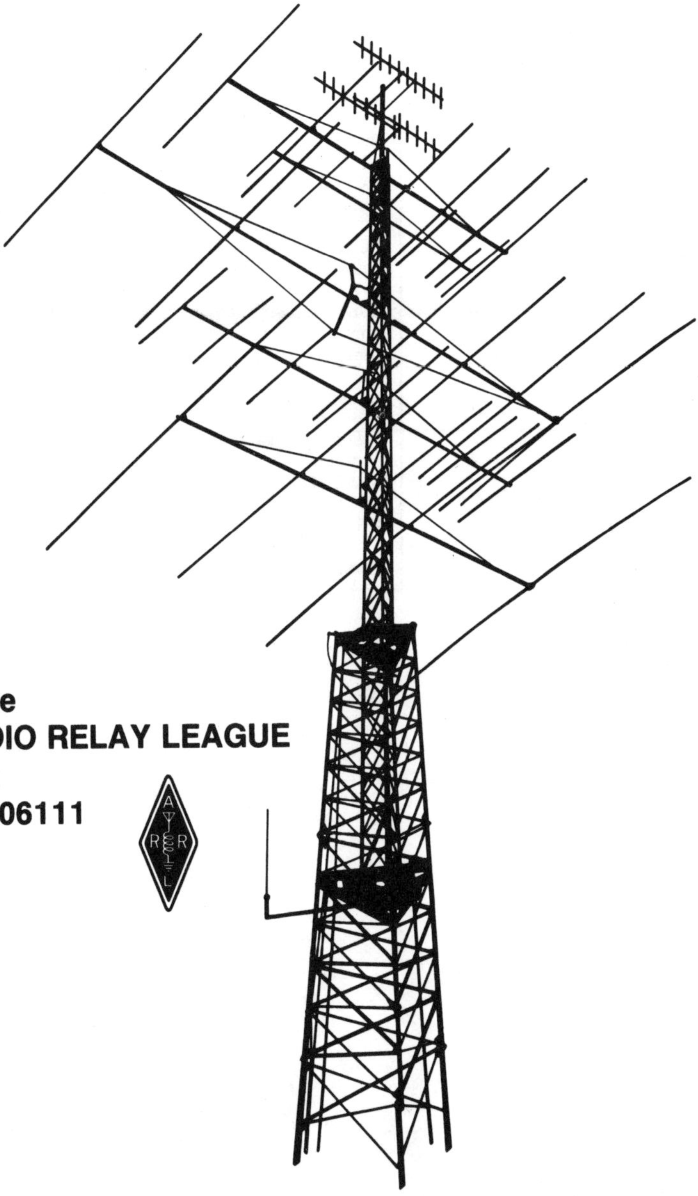

Published by the
AMERICAN RADIO RELAY LEAGUE
225 Main Street
Newington, CT 06111

Printed in USA

First Edition

$15.00 in USA and possessions

ISBN: 0-87259-041-0

FOREWORD

This book deals with a subject of interest to nearly every radio amateur, the Yagi-Uda antenna. All aspects of the design of high performance antenna systems are methodically presented. Among the topics addressed by *Yagi Antenna Design* are:

An efficient algorithm for calculating Yagi-Uda performance characteristics with a computer

Verification of theoretical calculations by comparison with measurements

Systematic examination of performance versus element lengths and spacings for two, three, four, and six element Yagis

How to establish a good free space design and then optimize that design for maximum front-to-back ratio

How to select antenna height and to compensate for the effect of ground on the free space performance, and how tilting the boom affects performance

The relative performance of loop elements versus dipole elements

Vertical stacking arrangements: effects of height, spacing, number of Yagis, phasing, and non-parallel booms

Procedures for scaling Yagi designs to change the operating frequency and account for a change in element diameter, and for adjusting element lengths to correct for element tapering and element-to-boom mounting fixture effects

Complete designs for 7, 14, 21, and 28 MHz Yagi-Uda antennas.

Although this book represents a valuable reference work for the radio amateur, rapid advances in personal computing have already made possible anayses which extend beyond that which was feasible at the time the book was prepared. Thus, Dr. Lawson's work no doubt will continue to be expanded and refined, as he would have wanted.

David Sumner, K1ZZ
Executive Vice President

Newington, Connecticut
December 1986

PREFACE

This book is a newly edited collection and an extension of a series of articles that I authored and published in *Ham Radio* during the early part of 1980. My objective was to develop design concepts for high-quality Yagi-Uda antenna systems for potential use by Amateur Radio operators in the lower-frequency communication bands where ionospheric refraction occurs; that is, frequencies between 5 and 30 MHz. The approach was to start with a reasonably good physical model of such an antenna system and compute its properites. Computation methodology was developed and adequately validated by comparison with a wide range of published experimental results. This methodology, implemented on a modern computer, allowed a quick and accurate investigation of the performance of a great many complex antenna systems. This happy circumstance makes possible a task that, experimentally, could never have been done or at best would require many lifetimes to perform. As a result, many brand-new conceptual ideas concerning these systems have emerged.

The series of articles in *Ham Radio* were constrained to meet the requirements of a magazine publication format and schedule; these limitations are removed in this book, so that a more coherent presentation is possible. The reader will note that I have made alterations to certain equations and numerical quantities in several places; these changes reflect the use of improved approximations but do not sensibly affect any of the results. In addition I have added, in Chapter 8, a number of representative designs with practical ways for the reader to use various element-taper schedules.

I am indebted to many who have written helpful comments to me regarding the original series. I also appreciate the efforts of the *Ham Radio* editors, who have made this publication possible.

I hope that this book will be useful and stimulating, not only to those who wish to build antennas, but to those who wish to improve the basic theoretical framework and computational techniques.

JAMES L. LAWSON, W2PV
Schenectady, New York
January, 1980

EDITORS' PREFACE

Jim completed this revision of his legendary series of *Ham Radio* articles in early 1980. For a number of reasons, the publishers of *Ham Radio* decided not to produce the manuscript as a book. Shortly before Jim's death in 1982, we volunteered to search for a means to publish Jim's work that would meet his exacting standards.

Obviously, this project has taken longer than we had hoped. Early on, Dave Sumner, K1ZZ, arranged for ARRL to sponsor the production of a camera-ready version. Bob Myers, W1XT, of Ethan Allen Printers, did much of the early typesetting. Craig Clark, N1ACH, of *Ham Radio* tracked down the original magazine artwork and generously offered the entire collection for the project. All of the radiation patterns were redrawn by computer to produce the ARRL polar plot format. (Small discrepancies between the plots and the values found in the tables are caused by a minor change in the models for self- and mutual resistance. Undiscovered major discrepancies are blunders entirely the responsibility of the first editor named below.) We must also acknowledge the considerable behind-the-scenes efforts of Dick Newell, AK1A.

The very large ratio of figures and tables to text has made the logical presentation of this material exceptionally challenging. The excellence of the outcome is the work of the ARRL production team: Debby Strzeszkowski, Shelly Chrisjohn, Sue Fagan and David Pingree. Sue Fagan designed the striking cover. Mark Wilson, AA2Z, coordinated the project at the ARRL end.

The surprising number and persistence of your enquiries regarding this project has sustained our enthusiasm. We believe that Jim would have been gratified by your interest and pleased by this outcome.

Bill Myers, K1GQ
Clarke Greene, K1JX
December 1986

ABOUT THE AUTHOR

Dr. James L. Lawson, W2PV was born in southern India of missionary parents. He was first licensed in 1934 as W9SSP while attending Kansas University. He went on to obtain his PhD in Physics from the University of Michigan, where he held the call sign W8QUI. From 1940 to 1945, he worked at the MIT Radiation Laboratory, where he made important contributions to the development of radar. Following the war, he joined the General Electric Research Laboratory (now the Research and Development Center), where he managed a number of key programs before his retirement in 1981. His technical specialties included nuclear and thermonuclear physics, particle accelerators, computers, and information systems.

CONTENTS

Chapter 1 **Performance Calculations**

Antenna Properties 1-2
Modeling 1-3
Computational Methodology 1-5
Element Self-Impedance 1-5
Mutual Impedance 1-9
Element Currents 1-9
Input Impedance and Directivity 1-11
Writing Computer Programs 1-12
Validation 1-15
NBS Yagi Experiments 1-15
Yagi Gain and Patterns 1-16
Effect of Director Length 1-22
Gain Variations 1-24
Comparison Summary 1-26

Chapter 2 **Simple Yagi Antennas**

Two-Element Beams 2-1
More Than Two Elements 2-7
Performance Characteristics 2-22
Element Illumination 2-24
Summary 2-31

Chapter 3 **Yagi Antenna Performance Optimization**

Parasite Length Variations 3-3
Parasite Placement Variations 3-11
Front-to-Back Optimization 3-19
Optimum Design 3-24
Design Example 3-24
Number of Reflectors 3-29
Missing Parasites 3-33
Summary 3-35

Chapter 4 **Loop Antennas**

Square Loop Model 4-1
Other Driven Loops 4-9
Multiloop Arrays 4-12
Summary 4-16

Chapter 5 **The Effects of Ground**

Reflections From A Plane Ground 5-1
Ground Curvature Effects 5-2
Image Models 5-3
Propagation Elevation Angles 5-4
Antenna Performance Over Ground 5-6
Best Height 5-12
Antenna Upward Tilt 5-14
Summary 5-15

Chapter 6 **Stacking**

Vertical Stacking Arrangements 6-2
Excitation 6-2
Two-Array Stack 6-3
Phase-Derived Fill 6-13
Three and Four-Array Stacks 6-15
Optimization of Stack Arrays 6-16
Orthogonal and Antiparallel Stacked Yagis 6-17
Summary 6-18

Chapter 7 **Practical Design**

Preferred Antenna Designs 7-1
Radius Scaling 7-3
Taper Corrections 7-5
Boom and Element Clamping Correction 7-11
Examples of Three-Element Beams 7-13
Summary 7-17

Chapter 8 **Practical Amateur Yagi Antennas**

Designs for 7.15 MHz 8-3
Designs for 14.2 MHz 8-4
Designs for 21.3 MHz 8-7
Designs for 28.5 MHz 8-8

CHAPTER 1

PERFORMANCE CALCULATIONS

Since its invention by H. Yagi and S. Uda of Tohoku University in 1926, the Yagi-Uda antenna—commonly referred to as the Yagi—has received a great deal of theoretical and experimental attention.[1] The Yagi has also become the most widely used Amateur Radio communications antenna, not only because of its excellent performance characteristics over the rather narrow frequency bands occupied by Amateurs, but also because of its remarkable tolerance to construction variations and even construction faults; it is an antenna that "wants to work."

For those interested in the theoretical basis of Yagi linear arrays, there are several excellent reference works available including those by Uda and Mushiake, Kraus, King, and Walkinshaw; these also contain many additional relevant references.[2,3,4,5] For those interested in the experimental side of Yagi arrays, attention is drawn to Ehrenspeck and Poehler, Lindsay, Greenblum, and Viezbicke.[6,7,8,9] Reference 9, National Bureau of Standards Technical Note 688, gives results of a decade-long NBS experimental investigation of Yagi antennas. It contains a wide range of experimental results, which can be used to test theoretical ideas.

Despite the extensive publications in existence on Yagi arrays, it appears that designers of high-performance Yagi antennas are faced with four facts:

1) Accurate antenna experiments are very difficult to make, especially if the antennas are designed to be used over ground. There are many variables. It is difficult, if not impossible, to avoid unwanted reflections. Also, accurate instrumentation is simply not available for many of the quantities to be measured—current distribution in the parasitic elements, for example.

2) Conceptual design ideas are often misleading; for example, the concept of "optimum element spacing." Optimum, indeed, but with respect to what?

3) Practical design of real antenna components in some cases does not exist; a physically "tapered" element, for example, must be significantly longer than an equivalent cylindrical element, but by how much?

4) A good theoretical basis for design is uncertain. A high-

[1]This and other footnotes are given at the end of this chapter

performance Yagi system is necessarily complex, with many adjustable parameters. Because of this complexity, a conceptual approach to superior design has not emerged. Moreover, the physical and mathematical simulation of such a system is potentially risky; not only must the physical model be valid, but the mathematical computations, which necessarily involve approximations, must be made with adequate accuracy.

The objectives of this book are to (1) consider all of these facts; (2) make significant contributions to both theoretical and practical bases for Yagi antenna design; (3) dispel misleading concepts; and (4) stimulate readers to make good, accurate, and relevant experiments. The subject of design is addressed primarily in terms of antenna arrays that are useful in the frequency range from 5 to 30 MHz, including designs ranging in size up to the largest practical levels. "Conventional" Yagi components are used, although there is some discussion of quads and of the quad-Yagi hybrid known as the quagi. For convenience, horizontal polarization is used unless otherwise stated.

Antenna Properties

Before beginning an investigation of antenna characteristics, it is necessary to discuss design criteria: What antenna properties are important, and how are these properties defined in quantitative terms? The antenna user is concerned with several properties:

1) Antenna gain
2) Pattern (including front-to-back ratio, *F/B*)
3) Bandwidth
4) Feed line matching, or standing-wave ratio
5) Cost
6) Longevity (wind survival, corrosion resistance)

Of these, the first four are electrical properties; the last two depend basically on construction engineering and are not discussed further.

Antenna gain, pattern, and bandwidth (items 1, 2, and 3) must be defined rather carefully. The gain is clearly of paramount importance; the definition of peak directivity is used for all situations.[10] *Directivity* in a specified direction is 4π times the ratio of the radiation intensity in that direction to the total power (P_T) radiated by the antenna. *Peak directivity* is the value of directivity in the direction of its maximum value.

An isotropic antenna has the same directivity in all directions; its radiation intensity in any direction is $P_T/4\pi$ and thus its peak directivity is 1 (0 dB). It has become common practice to refer antenna gain to the gain of this fictional antenna, which amounts to subtracting 0 dB from the directivity and calling the result *isotropic gain*, denoted by the symbol dBi. The (theoretical) peak directivity of a thin half-wavelength dipole is 1.64 (2.15 dB). Antenna gains which are referenced to a dipole (denoted by the symbol dBd) are converted to isotropic gain by adding 2.15 dB.

It is basically not practical to measure or specify the complete radiation pattern. The pattern characteristics of primary interest are the angular widths (horizontal and vertical) of the main beam and the amount of back radiation. The beam widths are rather simply related to the antenna gain (see Kraus, ref. 3, page 25). Radiation in the rear hemisphere is usually variable and consists of one or more lobes. Perhaps the single most meaningful measure of rear radiation is the *front-to-back ratio*, or *F/B*. This quantity is the ratio of peak directivity to directivity in the direction complementary to the direction of the peak. For example, if the peak directivity of a Yagi over ground occurs along the boom towards the directors at an elevation of 10 degrees above the horizon, then the complementary direction is 10 degrees above the horizon towards the reflector end of the boom.

Note that *F/B* is only one parameter of the complete antenna radiation pattern. It is perfectly possible, in principle, to have an antenna with a large *F/B* as defined above, but which has serious backward (but not directly back) lobes. Nevertheless this simple notion of *F/B* ratio is perhaps the most important single index of pattern behavior.

The third antenna property of interest is the *frequency bandwidth*, but there are at least three important bandwidths: (1) the bandwidth of the driving-point impedance (the electrical input); (2) the bandwidth over which the gain remains high; and (3) the bandwidth over which *F/B* remains high. All three are important, so it is necessary to display all three. This is best done by observing a quantitative frequency-swept plot of them all.

Feed line match is controlled by the matching system which transforms the antenna driving-point impedance to the transmission line characteristic impedance. The matching system contributes to (usually reduced) bandwidth in terms of standing-wave ratio on the transmission line. This effect depends on the details of the matching network and is beyond the scope of this book.

Antenna characteristics can be determined either by experimental measurements on a physical model, or by calculations from a mathematical representation of the physical model. Which should be used? Experimental measurements are laborious, and it is difficult to ensure accuracy. By contrast, computer calculations are fast and, in principle, can be made with great accuracy. Using a modern computer, a large number of antenna configurations can be investigated in a few days—a number it would take a lifetime to explore experimentally. Moreover, because of the inherent accuracy of calculations, subtle changes and radiative coupling effects can be explored. Therefore, it appears that if a computational procedure is believable, it can be used very effectively. Careful experimental tests are needed, however, to validate the computation methods.

Modeling

A real Yagi antenna can be represented by a set of parallel, cylindrical, conducting elements, each of which has space coordinates at its center of

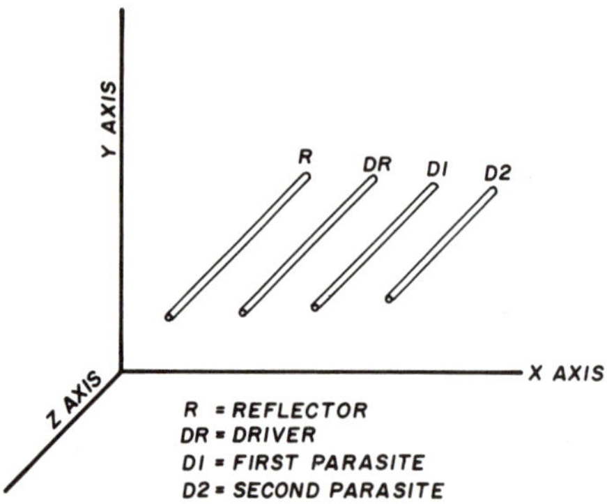

Fig. 1.1—Coordinate axes for the Yagi antenna model.

x, *y*, and *z*. As shown in Fig. 1.1 the Yagi is oriented so that the elements lie parallel to the z axis and the boom is parallel to the *x* axis. In free space, the origin of this coordinate system can be placed anywhere, but when modeling a Yagi over the ground it is advantageous to locate the origin on the conducting plane.

The elements themselves approximate a half-wave in length; the reflector is usually somewhat longer than a half-wavelength, the driven element is normally about a half-wavelength, and the directors are usually somewhat shorter than a half-wavelength. For convenience, and to make the representation independent of wavelength, all coordinates, lengths, and dimensions are expressed in wavelengths at some chosen design center frequency.

This representation is clearly a simplification of a real Yagi. A real Yagi, for example, does not ordinarily have strictly cylindrical elements; the real elements usually have telescoping diameters with connecting hardware clamps. Moreover, the elements are mounted at their centers with plates or brackets to the boom, which is usually a conductor.

The clamping and mounting hardware can be accounted for by correcting the *actual length* of the element to an *equivalent length*. Similarly a telescoping element will act as a cylindrical element of the same *average radius* and an equivalent length different from the actual length. The way in which the actual element dimensions can be converted to a equivalent cylindrical element is discussed in Chapter 7, as are corrections to length caused by mounting hardware. As a side note, the element radius taper corrections to convert actual lengths to equivalent cylinder lengths are substantial; this is often overlooked by builders who try to use someone else's element length schedule with different element tubing diameters.

The simplified Yagi model also neglects the conducting boom; this is justified only if the real Yagi is completely symmetric around the boom. Symmetry guarantees zero electrical potential at the center of each parasitic element and no mutual coupling to the boom. Hence no current will be induced in the boom by the parasitic elements. If the driven element is open at its center and driven from a balanced source, then it will induce no current along the boom. Thus, for the symmetric Yagi with balanced feed there are no currents in the boom, and therefore no effect of the boom on antenna performance.

Unbalanced feed systems such as the gamma match are usually not especially troublesome, because if the driving-point impedance of the element is relatively low, as it usually is, the voltage impressed onto the boom is also low, and boom current will be correspondingly low. Moreover, the loaded Q of the driven element is usually high enough to ensure reasonable symmetry of element currents; this also makes for low induced boom currents. Incidentally, it is possible to construct the real Yagi with insulating element-to-boom supports; this helps to ensure negligible boom currents. In any case, small corrections to element length due to the proximate boom are discussed in Chapter 7.

In the mathematical model one can arrange any number of parasitic elements and any number of drivers. Chapter 4 shows how to use the model to approximate a quad or a quagi. The model can easily be extended to include a "broadband" drive (as in the KLM antennas), where there is a main driver and one or more dependent drivers which are connected through a transmission line to the main driver.

Computational Methodology

With this conceptual model we are in a position to compute the performance of the Yagi array—to find out how the performance of the array varies with frequency near the design frequency. The first task is to compute the complex currents which flow in all elements as a result of driver excitation; to do this, it is necessary to determine both the self- and mutual impedances of all elements.

Element Self-Impedance

The self-impedance of a single nearly half-wavelength element in free space has been calculated by many authors; an excellent comparison of the various methods is given by Kraus (ref. 3, pages 272-276). Uda and Mushiake (ref. 2) use the method originated by Hallen (solution of the boundary-value problem), and present an approximate equation and a table which show the self-impedance of a cylindrical, nearly half-wavelength dipole as functions of radius and length.[11] It is apparent from the table that a half-wavelength dipole has an impedance of about 73 + j40 ohms, so a somewhat shortened antenna is needed to resonate, that is, to show zero reactance. The required shortening is basically only a function of K, which is defined as the ratio of wavelength to element radius. In Table 1.1, I extended Uda and Mushiake's reactance data to a wider range of values, and calculated the reactance of a full halfwave dipole, $X_{0.5}$, and of an element 0.45λ long, $X_{0.45}$. Also tabulated is the change in reactance when shortening the element from element from 0.5λ to 0.45λ, ΔX.

The reactance of the halfwave dipole can be easily approximated by a limited power series in log K:

$$X_{0.5} \approx 33.25 + 3.19 \log K - 0.35 (\log K)^2 \qquad (1.1)$$

Table 1.1

Cylindrical dipole self-reactance (ohms) for different wavelength to radius ratios (K). $X_{0.5}$ is the reactance of a dipole element 0.5 wavelength long, $X_{0.45}$ is the reactance of a dipole 0.45 wavelength long and ΔX is the reactance change when shortening the element from 0.5 wavelength to 0.45 wavelength.

K	$X_{0.5}$ (Ref.2)	$X_{0.5}$ (1.1)	$X_{0.5}$ (1.2)	$X_{0.45}$ (Ref. 2)	ΔX	ΔX (1.3)
1×10^1	34.2	36.1	34.2	23.1	−11.1	−9.2
3×10^1	36.7	37.2	36.6	6.4	−30.3	−29.8
1×10^2	38.2	38.2	38.2	−14.1	−52.3	−52.3
3×10^2	39.0	39.0	39.1	−33.6	−72.6	−72.9
1×10^3	39.6	39.7	39.6	−55.5	−95.1	−95.4
3×10^3	40.0	40.1	40.0	−75.7	−115.7	−116.0
1×10^4	40.4	40.4	40.4	−98.1	−138.5	−138.5
3×10^4	40.6	40.5	40.6	−118.6	−159.2	−159.1
1×10^5	40.8	40.5	40.8	−141.1	−181.9	−181.6
3×10^5	41.0	40.2	40.6	−161.8	−202.8	−202.2
1×10^6	41.1	39.8	39.9	−184.4	−225.5	−224.7

or

$$X_{0.5} \approx 24.1591 + 14.5151 \log K - 5.3677 (\log K)^2 + 0.9352 (\log K)^3 - 0.06184 (\log K)^4 \tag{1.2}$$

Equation 1.1 is a three-term approximation, and (1.2) is a more accurate five-term approximation. Either one is satisfactory in the chief region of interest ($10^2 < K < 10^4$). The quality of both approximations is shown in Table 1.1.

A plot of ΔX as a function of log K (Fig. 1.2) shows that the calculated points lie remarkably close to a straight line. Thus ΔX can be represented quite well by a simple empirical expression:

$$\Delta X \equiv X_{0.45} - X_{0.5} = -43.1 \log K + 33.9 \tag{1.3}$$

The quality of this approximation is also shown in Table 1.1.

From Uda and Mushiake's table it is apparent that in the region of interest (length around a half-wavelength) the reactance of any cylinder is essentially linearly related to length; thus for any element length, l, expressed in wavelengths the reactance is approximately:

$$X_l = X_{0.5} + 20\,\Delta X\,(0.5 - l) \tag{1.4}$$

Reactance vanishes at the resonant length, l_R, also expressed in wavelengths:

$$X_{l_R} = 0 = X_{0.5} + 20\,\Delta X\,(0.5 - l_R) \tag{1.5}$$

Then:

$$l_R = 0.5 + X_{0.5} / 20\,\Delta X \tag{1.6}$$

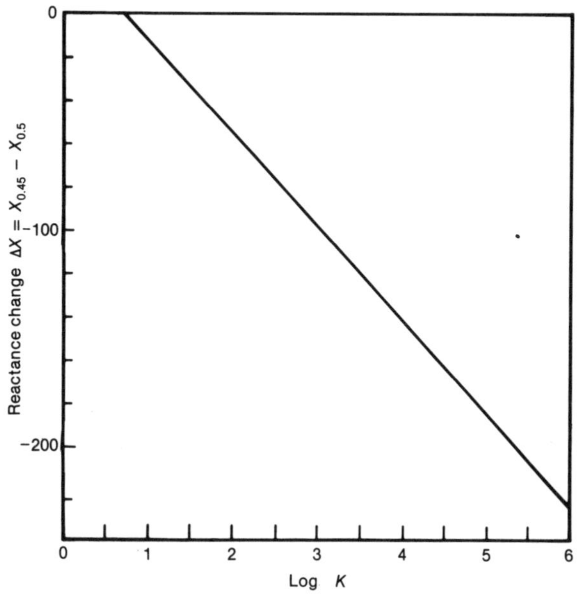

Fig. 1.2—Differential cylindrical dipole self-reactance vs. wavelength-to-radius ratio.

Table 1.2

Ratio of cylindrical dipole resonant length (l_R) to a half-wavelength for different wavelength to radius ratios (K).

K	$2l_R$
3×10^1	0.877
1×10^2	0.927
3×10^2	0.946
1×10^3	0.958
3×10^3	0.966
1×10^4	0.971
3×10^4	0.974
1×10^5	0.978
3×10^5	0.980
1×10^6	0.982

The resonant length is always shorter than a half-wavelength; it is interesting to quantify the difference. Table 1.2 shows calculations of the ratio of l_R to a half-wavelength computed from (1.6), (1.3) and (1.2) as a function of K. A corresponding plot is shown in Fig. 1.3. Note that the resonant element is substantially shorter than a half-wavelength for thick cylinders (low values of K) but even for extremely thin wires (high values of K) significant shortening is required.

Using (1.4) and (1.5) the reactance of an element of length l wavelengths is expressed as:

$$X_l = -20\,\Delta X\,(l - l_R) \qquad (1.7)$$

where all variables are wavelength (or frequency) dependent. In investigating frequency displays of element properties it is awkward to use these changing variables.

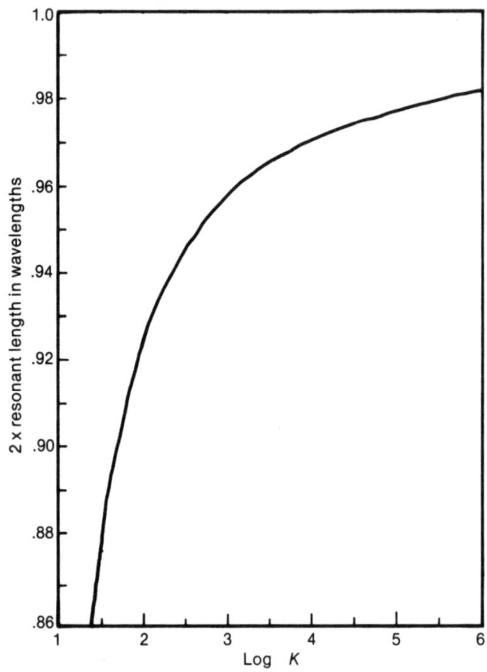

Fig. 1.3—Ratio of cylindrical dipole resonant length to a half wavelength vs. wavelength-to-radius ratio.

Therefore define:

f = frequency
f_D = design frequency
F = f/f_D = normalized frequency variable
λ_D = wavelength at design frequency
K_D = ratio of λ_D to element radius
l_D = ratio of element length to λ_D Then:

$$K = K_D / F \tag{1.8}$$

and

$$l = l_D F \tag{1.9}$$

where K_D and l_D are independent of the frequency scan variable, F. Now (1.7) may be rewritten:

$$X_l = A(F - F_R) \tag{1.10}$$

where

$$A = -20 \, \Delta X \, l_D \tag{1.11}$$

and

$$F_R = l_R / l_D \tag{1.12}$$

Equation 1.12 can be easily derived from (1.6), (1.1) or (1.2), (1.3) and (1.8). The constant A in (1.11) is, of course, the slope of reactance with normalized frequency. Recall that a simple series RLC circuit in the neighborhood of resonance displays an input reactance of:

$$X = RQ\,(F / F_R - F_R / F) \approx 2RQ\,(F - F_R) \tag{1.13}$$

where Q is the electrical Q factor and R is the series resistance. A comparison of (1.13) and (1.10) indicates that electrically the dipole element behaves essentially like the lumped circuit and that the quantity $A/2R$ can be equated to circuit Q. The Q of the dipole element is an important parameter basically determining bandwidth.

The self-impedance of the element is thus:

$$Z_l = 73 + jX_l \tag{1.14}$$

The accuracy of this expression should be good to a very few percent for elements within a few percent of a half-wavelength long.

Mutual Impedance

Now we must consider the mutual impedance between two nearly half-wavelength elements separated by a distance of s measured in wavelengths. Several authors have made calculations of the real and imaginary components of this complex quantity for the limiting case of infinitely thin, half-wavelength elements. Exponential integral equations, plots, and tables are shown in Kraus (ref. 3, pages 265-268), and Uda-Mushiake (ref. 2, pages 69-70). Kraus (page 266) also shows calculations by Tai for two cases of thicker, half-wavelength dipole elements. Tai's calculations suggest that inaccuracies caused by using the limiting thin case are not very large; therefore, for convenience, I have used it for all calculations. I have extended the table of Uda-Mushiake; Table 1.3 shows values in ohms for the real part of the mutual impedance (R_{MUT}) and for the imaginary part (X_{MUT}), as a function of element separation. For separations greater than $s = 2\lambda$, a reasonable approximation can be derived from:

$$R_{MUT} = \frac{19.06}{s} \sin 2\pi s, \quad s > 2\lambda \tag{1.15}$$

$$X_{MUT} = \frac{19.06}{s} \cos 2\pi s, \quad s > 2\lambda \tag{1.16}$$

Element Currents

Although some caveats are necessary, we now have the necessary tools to

Table 1.3

Cylindrical dipole mutual impedance for values of element spacing(s) between zero and two wavelengths.

s (λ)	R_{MUT} (ohms)	X_{MUT} (ohms)	s (λ)	R_{MUT} (ohms)	X_{MUT} (ohms)
0.00	73.1	42.5			
0.05	71.7	24.3	1.05	8.8	15.0
0.10	67.3	7.5	1.10	12.3	11.2
0.15	60.4	−7.1	1.15	14.5	6.7
0.20	51.4	−19.2	1.20	15.3	1.9
0.25	40.8	−28.4	1.25	14.6	−2.7
0.30	29.3	−34.4	1.30	12.6	−6.7
0.35	17.5	−37.4	1.35	9.6	−9.8
0.40	6.2	−37.4	1.40	6.0	−11.9
0.45	−4.0	−34.8	1.45	2.0	−12.7
0.50	−12.5	−29.9	1.50	−1.9	−12.3
0.55	−19.1	−23.4	1.55	−5.4	−10.8
0.60	−23.3	−15.9	1.60	−8.2	−8.4
0.65	−25.2	−7.9	1.65	−10.0	−5.3
0.70	−24.9	−0.3	1.70	−10.9	−2.0
0.75	−22.5	6.6	1.75	−10.6	1.4
0.80	−18.5	12.3	1.80	−9.4	4.5
0.85	−13.3	16.3	1.85	7.4	7.0
0.90	−7.5	18.6	1.90	−4.8	8.7
0.95	−1.6	19.0	1.95	−1.9	9.5
1.00	4.0	17.7	2.00	1.1	9.4

calculate the parasitic element currents. Recall that the physical model of the Yagi is a good representation only to the extent that proper corrections can be made for element hardware variances (clamping and mounting hardware and element radius taper). These corrections are discussed in Chapter 7. Also recall that the computation of self- and mutual impedances are approximations.[12] Though they are probably good approximations, and should give reasonably accurate results, one should not rely on them for accuracy better than a few percent.

The first step is to calculate the complex currents (or magnitudes and phases) of all parasitic elements given the currents or voltages applied to all drivers. The method is uncomplicated, following the technique of P.S. Carter shown in Kraus (ref. 3, page 302).[13] For simplicity, I illustrate how this is done using one driver and three parasitic elements; extension to any number of elements is obvious. For each element add all voltages and equate to the terminal voltage V_n:

$$
\begin{aligned}
I_1 Z_{11} + I_2 Z_{12} + I_3 Z_{13} + I_4 Z_{14} &= V_1 \\
I_1 Z_{21} + I_2 Z_{22} + I_3 Z_{23} + I_4 Z_{24} &= V_2 \\
I_1 Z_{31} + I_2 Z_{32} + I_3 Z_{33} + I_4 Z_{34} &= V_3 \\
I_1 Z_{41} + I_2 Z_{42} + I_3 Z_{43} + I_4 Z_{44} &= V_4
\end{aligned}
\tag{1.17}
$$

Assume in this example that the first three elements are parasites, i.e., $V_1 = V_2 = V_3 = 0$ and that the fourth element is driven with the complex voltage V_4. Z_{nn} is recognized as the complex self-impedance of the nth element, and Z_{jk} (which is the same as Z_{kj}) is the mutual impedance between the jth and kth element. Thus, all of these impedances can be calculated once the positions (and hence separations) of the elements are specified. Since there are four linear equations with four unknowns, I_1, I_2, I_3, and I_4, the easiest way to solve this array is by a matrix inversion. In matrix notation (1.17) is represented by:

$$ZI = V \tag{1.18}$$

where Z is a 4×4 matrix, I and V are four element vectors, and all components are complex numbers. The solution is

$$I = Z^{-1} V \tag{1.19}$$

where Z^{-1} is the matrix inverse of Z.

The process of complex matrix inversion is readily accomplished with a computer, with a program usually called CLINEQ under FORTRAN IV. Although the actual solution is usually done through a mathematical process known as Gaussian elimination, the result is equivalent to matrix inversion. With this technique, a computer will provide solutions quickly for very large arrays of fifty elements or more.

If one wishes to specify the driven element current, I_4, instead of voltage, V_4, rewrite the first three parasitic equations of (1.17) as:

$$\begin{aligned} I_1 Z_{11} + I_2 Z_{12} + I_3 Z_{12} &= -I_4 Z_{14} \\ I_1 Z_{21} + I_2 Z_{22} + I_3 Z_{22} &= -I_4 Z_{24} \\ I_1 Z_{31} + I_2 Z_{32} + I_3 Z_{32} &= -I_4 Z_{34} \end{aligned} \tag{1.20}$$

This (smaller) array can be solved in the same way for I_1, I_2, and I_3 and the results used in the fourth equation:

$$I_1 Z_{41} + I_2 Z_{42} + I_3 Z_{42} + I_4 Z_{44} = V_4 \tag{1.21}$$

to solve for V_4.

Input Impedance and Directivity

Once the complex currents in each element are known, the Yagi properties of interest can be computed by straight forward formulas. First, the input impedance is found from:

$$Z_{IN} = V_{DR}/I_{DR} \tag{1.22}$$

where V_{DR} and I_{DR} are the complex voltage and current for the driven element (V_4 and I_4 in the example of the previous section). The power input to the antenna is then:

$$P_{IN} = \tfrac{1}{2} \left| I_{DR} \right|^2 R_{IN} \tag{1.23}$$

where R_{IN} is the real part of the input impedance.

Two assumptions are required to compute the directivity: (1) all of the input power is radiated; and (2) the amplitude of the current in each element varies sinusoidally from maximum at the element center to zero at the tips. The first assumption is reasonable for HF Yagis constructed with aluminum tubing, and the second is quite accurate for element lengths near λ/2. The general expression for directivity is:

$$D = \frac{60}{P_{IN}} \left| \sum_{i=1}^{n} e^{j2\pi r_i F} I_i g_i \right|^2 \tag{1.24}$$

where F is the normalized frequency, I_i is the complex current at the center of the ith element, and n is the number of elements. For the H-plane (that is, the xy plane in Fig. 1.1):

$$r_i = x_i \cos\phi + y_i \sin\phi \tag{1.25}$$

and

$$g_i = 1 - \cos\pi\, l_i \tag{1.26}$$

where x_i and y_i are the coordinates of the center of the ith element, l_i is the length of the ith element in wavelengths, and ϕ (the elevation angle) varies from 0 degrees along the x axis through 90 degrees along the y axis to 180 degrees along the $-x$ axis. Note that x and y are specified in terms of wavelength *at the design frequency*. For the E-plane (the xz plane):

$$r_i = x_i \sin\theta \tag{1.27}$$

and

$$g_i = [\cos(\pi l_i \cos\theta) - \cos\pi l_i] / \sin\theta \tag{1.28}$$

with θ (the azimuth angle) equal to 0 along the z axis, 90 degrees along the x axis and -90 degrees along the $-x$ axis. Note that (1.28) contains a division by zero when $\theta = 0$ degrees; in fact the pattern factor g_i approaches zero as θ approaches zero (corresponding to the well-known null off the ends of dipole elements). Therefore in the computer program set $g_i = 0$ when $\theta = 0$.

Writing Computer Programs

Specific computer programs for calculating array properties are not hard to write, but the details do depend on the specific computer and available software libraries and subroutines. This section explains the information flow used in FORTRAN programs for computing element currents and Yagi performance. I first create a labeled input file containing all the necessary information about the Yagi. The label itself is usually an abbreviated reminder of the particular Yagi. This input file contains (1) a statement of the number of parasites and the number of drivers; (2) information on each parasite, i.e., x and y coordinates, length and radius, all specified in terms of wavelength at a given design frequency; and (3) information on each driver, i.e., x and y coordinates, length and radius, and driving-point voltage and phase.

The element-current program first asks for the input data file label and the frequencies (in terms of the center design frequency) at which the computations are to be made. With this starting information, the input file is called and read. For each frequency specified, a computation is made to determine the (complex) values of all terms in the Z matrix; self-impedances are calculated from (1.14), and mutual impedances are given by an interpolation routine from the data in Table 1.3 or from the power series approximations (1.15) and (1.16).[14]

Once the matrix is complete and the voltage vector noted from the input file, matrix inversion is accomplished; this generates the individual element complex current solutions. The entire result is then written into an output file for later use. The output file contains (1) a statement of frequency for each computation; (2) the number of parasites and number of drivers; and (3) for each element, its x and y coordinates, its resonant frequency F_R, the parameters $X_{0.5}$, ΔX, l_R, K, and A, and the magnitude and phase of its current. The element-current program continues to compute until all initially specified frequencies are satisfied.

This output file is an input file to other computer programs which are designed to produce displays of interesting Yagi properties. One such program calculates and plots the radiation patterns for the H-plane and the E-plane. This is done by reading the element current file, and for a specified sequence of angles (say every 1 or 2 degrees), computing the directivity from (1.24). The radiation patterns are calculated for each frequency (read from the new input file) and can be easily plotted by one of the plot routines. It is convenient to also display the driving-point impedance, the peak directivity value found in the elevation angle search (usually in 1 degree intervals), the angle at which it occurs, and the front-to-back ratio.

Another program of great utility is one which reads the element current file and computes and displays a number of useful Yagi properties. These are peak directivity and the elevation angle at which it occurs, reverse directivity at the same elevation angle, front-to-back ratio, and the driving-point impedance. Since this program does not spend time displaying the complete patterns, it is fast and therefore capable of rapidly running through a large number of different situations.

Programs can be written that will perform equivalent calculations for differ-

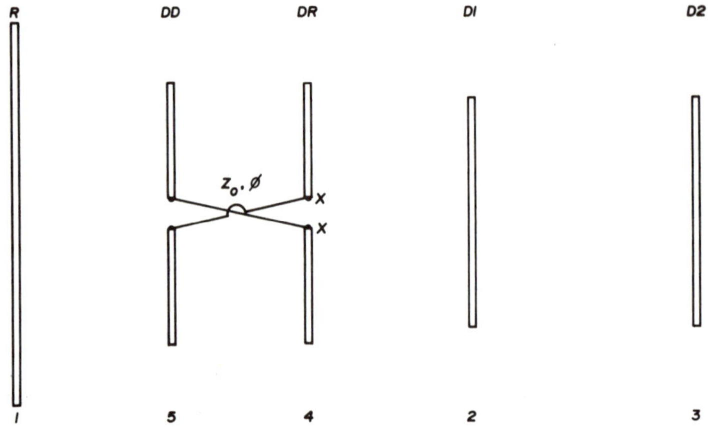

Fig. 1.4—Layout of a Yagi with a broadband feed system of the type manufactured by KLM.

ent antenna systems. As an example, one can write a program to handle a broadband drive system as used in the KLM antennas, but to do this requires a little manipulation. In this type of antenna the drive system is not just a single element but actually consists of a voltage-fed main driver in parallel with a dependent driver through a crossed transmission line with characteristic impedance Z_0 and phase delay angle ϕ. With conventional transmission line equations one can relate voltages and currents at both ends of this transmission line.

As an illustration, a five-element KLM-style beam can be simulated as shown in Fig. 1.4. The parasites are elements 1, 2, and 3, the main or master driver is 4, and the dependent or slave driver is 5 (connected through the transposed transmission line). Start with the five linear equations:

$$\begin{aligned} I_1 Z_{11} + I_2 Z_{12} + I_3 Z_{13} + I_4 Z_{14} + I_5 Z_{15} &= 0 \\ I_1 Z_{21} + I_2 Z_{22} + I_3 Z_{23} + I_4 Z_{24} + I_5 Z_{25} &= 0 \\ I_1 Z_{31} + I_2 Z_{32} + I_3 Z_{33} + I_4 Z_{34} + I_5 Z_{35} &= 0 \\ I_1 Z_{41} + I_2 Z_{42} + I_3 Z_{43} + I_4 Z_{44} + I_5 Z_{45} &= V_4 \\ I_1 Z_{51} + I_2 Z_{52} + I_3 Z_{53} + I_4 Z_{54} + I_5 Z_{55} &= V_5 \end{aligned} \tag{1.29}$$

V_4 (the main driver voltage) is given, but V_5 (the dependent driver voltage) must be determined from the transmission line equations which relate V_5, I_5, V_4 and transmission line current at terminals xx. For a transposed lossless transmission line:

$$V_5 = -V_4 \sec\phi - jZ_0 I_5 \tan\phi \tag{1.30}$$

so that the final five linear equations become:

$$I_1 Z_{11} + I_2 Z_{12} + I_3 Z_{13} + I_4 Z_{14} + I_5 Z_{15} = 0$$

$$I_1 Z_{21} + I_2 Z_{22} + I_3 Z_{23} + I_4 Z_{24} + I_5 Z_{25} = 0$$

$$I_1 Z_{31} + I_2 Z_{32} + I_3 Z_{33} + I_4 Z_{34} + I_5 Z_{35} = 0$$

$$I_1 Z_{41} + I_2 Z_{42} + I_3 Z_{43} + I_4 Z_{44} + I_5 Z_{45} = V_4$$

$$I_1 Z_{51} + I_2 Z_{52} + I_3 Z_{53} + I_4 Z_{54} + I_5 (Z_{55} + jZ_0 \tan\phi) = -V_4 \sec\phi \tag{1.31}$$

Note that the input file for the antenna must contain information about the number of parasites, the number of independent or master drivers, and the number of dependent or slave drivers; the dependent or slave driver must contain a statement as to which is its master driver, and must provide Z_0 and ϕ for the transposed transmission line. The Z matrix and V vector must be modified as shown, then the program can proceed as before.

Validation

With the tools for computing a wide spectrum of antenna characteristics at hand, it is crucial to ask for some experimental validation. I have already commented on the inaccuracies inherent in physical modeling (although such inaccuracies are believed to be of little consequence), and have discussed approximations which have been used in various computations. These approximations are expected to be most serious for those antenna properties which depend critically on precise element currents, such as F/B ratio. On the other hand, it should be possible to calculate other properties such as directivity with reasonable accuracy (comparable to the accuracy of the currents themselves). The calculations are expected to produce superior results for short antennas (few elements) and for frequency regions relatively close to the design center frequency.

NBS Yagi Experiments

Experimental information suitable for validation is rather sparse. While a great amount of experimental information has been published, it is difficult to find examples of accurate measurements made under conditions where external factors have been properly considered or eliminated.

For example, let's examine the experimental results reported by Peter Viezbicke in NBS Technical Note 688 (ref. 9). This publication provides a rich range of gain and pattern measurements for a wide variety of Yagi configurations. In most cases, all relevant dimensions of the Yagi design are given, so this publication is a fine vehicle to test the validity of theoretical computational methodology; shortly, I will attempt to make such comparisons.

I must first comment on the NBS experimental approach. Viezbicke states that all tests on the (receiving) test antennas were carried out using a nonconducting Plexiglas boom mounted three wavelengths (3λ) above ground (although readers of the NBS paper will note that the first figure shows 2λ). The (transmitting) generator antenna was mounted 320 meters (about 430λ) away and at a height above ground to illuminate the receiving antenna at "grazing angles." I interpret this to mean that the transmitter is also at a height over ground of 3λ. The nature of the intervening ground, which is highly relevant, is not described. A reference half-wavelength dipole is mounted about 5λ to the side of the antenna under test (also at 3λ above ground). All tests were made using a frequency of "400 MHz" but the frequency precision is never mentioned, and this may be quite important, as shall be seen.

If my interpretation of the experimental setup is correct, received field strength will be the vectorial sum of the field from the direct ray and the field from the ray reflected from the ground at a point midway between transmitter and receiver. At grazing angles, the reflectivity of the ground is near unity and for horizontally polarized radiation ground reflection will give a phase change of 180 degrees. Thus, the two rays interfere and nearly cancel each other, making the received energy nearly zero. It can be seen that the actual received energy is extremely sensitive to the nature of ground midway between transmitter and receiver, i.e., its true reflection coefficient, its height, and its degree of flatness (which influences its focusing properties). Moreover, since the reference antenna "sees" a different ground patch midway to the transmitter, its inherent sensitivity can be different from that of the antenna under test. For example, if the ground patch seen by the reference antenna is only 1 foot higher than the ground patch seen by the test antenna, a systematic error of about 2 dB will occur. It is clear, therefore, that the NBS experimental setup invites systematic errors.

Yagi Gain and Patterns

Let me turn now to the experimental results reported by Viezbicke. I will start with the measured gain of a two-element (dipole and reflector) Yagi as a function of spacing. Unfortunately, Viezbicke failed to state an essential dimension—the actual length of the reflector. On the assumption that it was 0.482λ (which was usually used for the other Yagis), I calculated the theoretical result shown in Fig. 1.5. The theoretical gain is expressed in dBi; a direct comparison can be made by adding 2.15 dB to the experimental NBS findings.

The comparison in Fig. 1.5 is totally unsatisfactory; not only are the shapes of the plots different, the peak gains are very different. At a spacing of 0.2λ, Viezbicke measured 4.77 dBi, whereas I calculated 6.70 dBi. In an attempt to resolve this conflict, I numerically summed and properly averaged the measured pattern of the same two-element Yagi (Viezbicke's Fig. 13). I found the directivity, or isotropic gain, to be 6.71 dBi. Thus, the measured pattern exhibits gains which agrees with my theoretical result. I can only conclude

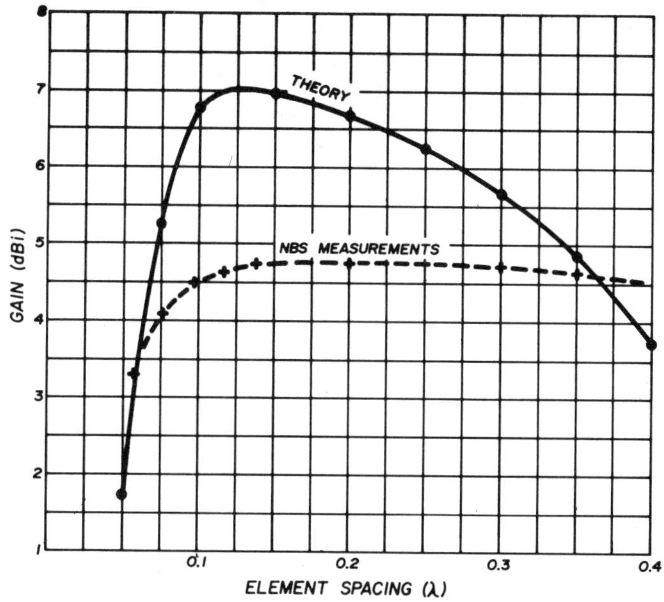

Fig. 1.5—Gain of a driven element and reflector for different spacings between elements (from NBS Report 688, Fig. 1).

that some unexplained error was present for the NBS series of measurements shown in their Fig. 1.

In the NBS paper a set of six specific Yagi designs is shown ranging in overall boom length from 0.4 to 4.2 wavelengths. If you refer directly to the NBS publication you will note that the first director for the 0.4λ three-element design is shown in their Fig. 9 to be 0.442λ long rather than 0.424λ, as shown in their Table 1. With this correction, the specifications for the six NBS Yagi designs are listed here in Table 1.4.

I calculated the theoretical gains for these six Yagi designs, and also their theoretical patterns for comparison with the measured patterns shown in the NBS report. For convenience in making comparisons, the NBS patterns and my calculated theoretical patterns are shown side by side in Figs. 1.6 through 1.12. It is apparent that for all cases there is striking similarity, not only in the qualitative details of lobe structure, but also in most quantitative aspects. Careful scrutiny of the experimental results show some variances between the two halves; theoretically, of course, the two halves are totally symmetrical. Agreement of this kind is gratifying and demonstrates that the experimental patterns were made with great care and also that the theoretical calculations give valid answers. This is especially comforting since the 4.2λ Yagi design is very long and contains many parasites—precisely the situation where theoretical approximations (cylindrical element resonant lengths, mutual, and self-impedances) are most sensitive.

Table 1.4

NBS optimized lengths of parasitic elements for Yagi antennas of six different boom lengths (reflector spaced 0.2 wavelength behind the driven element, element diameter 0.0085 wavelength).

Boom length (wavelength)	0.4	0.8	1.2	2.2	3.2	4.2
Number of elements	3	5	6	12	17	15
Reflector length (wavelength)	0.482	0.482	0.482	0.482	0.482	0.475
Director length (wavelength)						
1st	0.442	0.428	0.428	0.432	0.428	0.424
2nd		0.424	0.420	0.415	0.420	0.424
3rd		0.428	0.420	0.407	0.407	0.420
4th			0.428	0.398	0.398	0.407
5th				0.390	0.394	0.403
6th				0.390	0.390	0.398
7th				0.390	0.386	0.394
8th				0.390	0.386	0.390
9th				0.398	0.386	0.390
10th				0.407	0.386	0.390
11th					0.386	0.390
12th					0.386	0.390
13th					0.386	0.390
14th					0.386	
15th					0.386	
Director spacing (wavelength)	0.200	0.200	0.250	0.200	0.200	0.308
Gain (dBd)	7.1	9.2	10.2	12.25	13.4	14.2
(dBi)	9.25	11.35	12.35	14.40	15.55	16.35

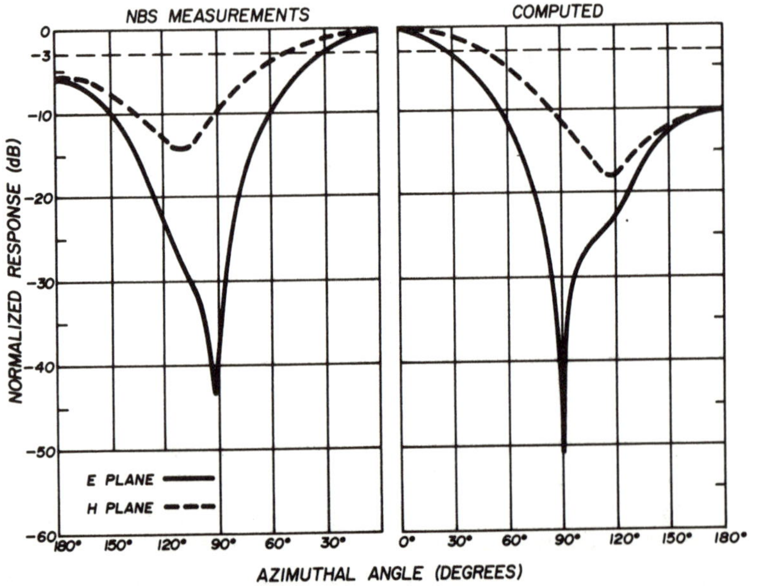

Fig. 1.6—Measured and computed *E*- and *H*-plane radiation patterns for a two-element Yagi with a boom length of 0.2 wavelengths (from NBS Report 688 Fig. 13).

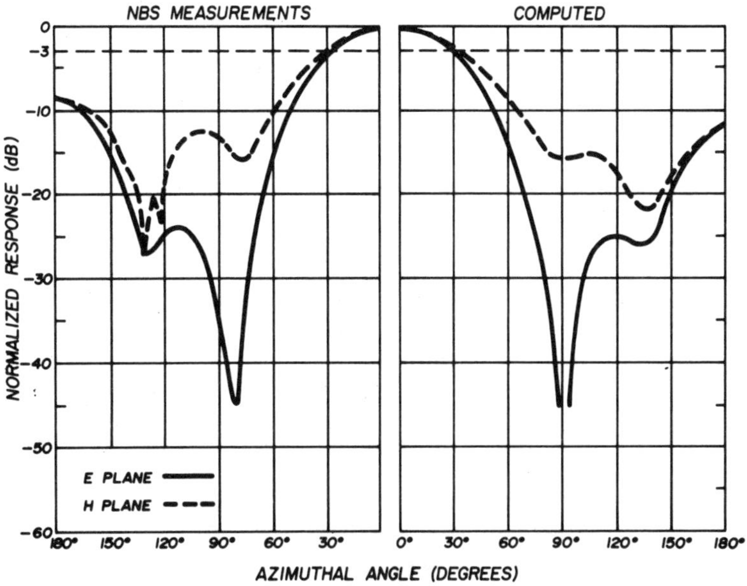

Fig. 1.7—Measured and computed *E*- and *H*-plane radiation patterns for a three-element Yagi with a boom length of 0.4 wavelength (from NBS Report 688 Fig. 14).

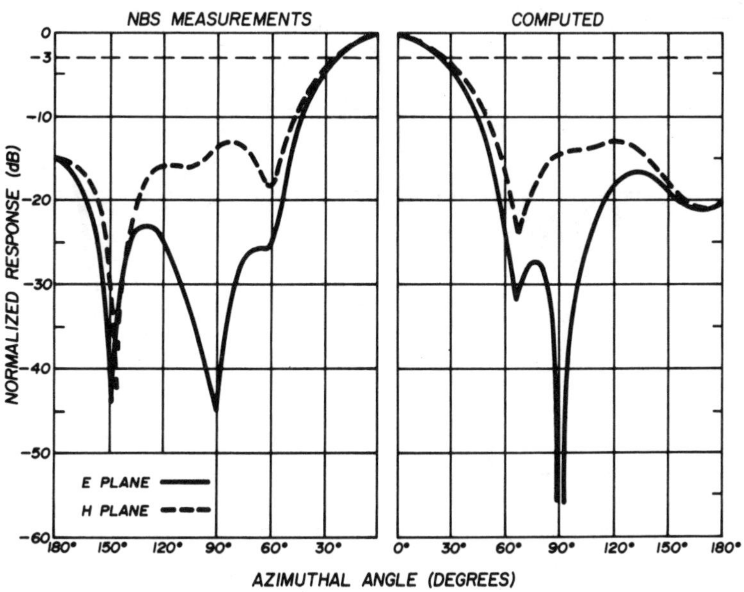

Fig. 1.8—Measured and computed *E*- and *H*-plane radiation patterns for a five-element Yagi with a boom length of 0.8 wavelength (from NBS Report 688 Fig. 15).

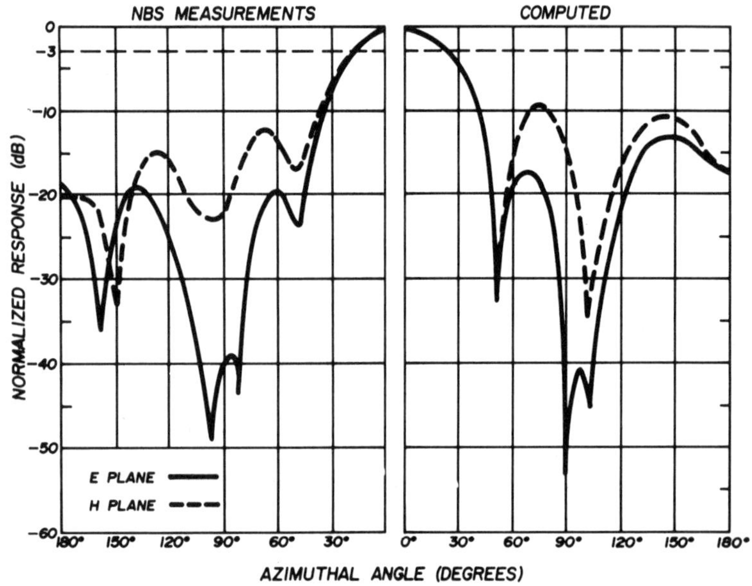

Fig. 1.9—Measured and computed *E*- and *H*-plane radiation patterns for a six-element Yagi with a boom length of 1.2 wavelengths (from NBS Report 688 Fig. 16).

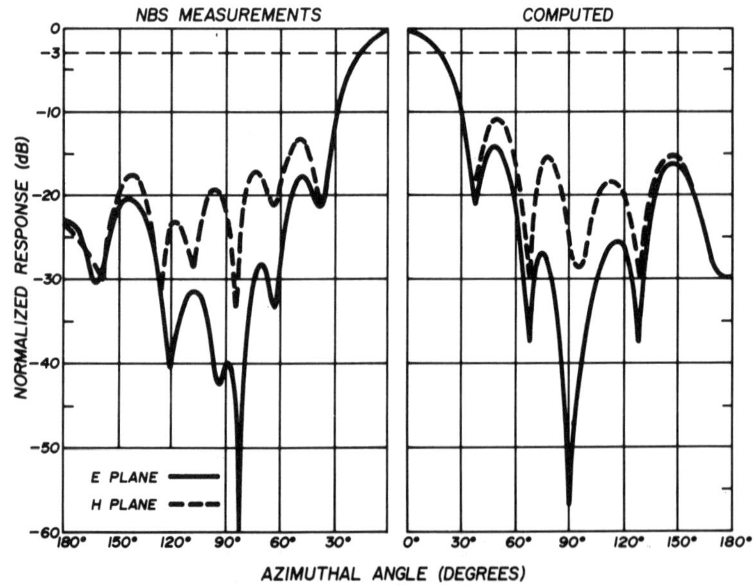

Fig. 1.10—Measured and computed *E*- and *H*-plane radiation patterns for a twelve-element Yagi with a boom length of 2.2 wavelengths (from NBS Report 688 Fig. 17).

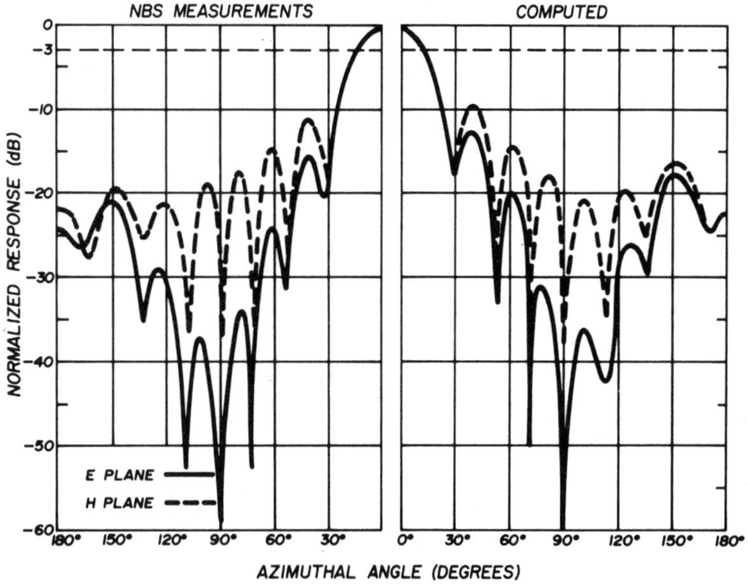

Fig. 1.11—Measured and computed *E*- and *H*-plane radiation patterns for a seventeen-element Yagi with a boom length of 3.2 wavelengths (from NBS Report 688 Fig. 18).

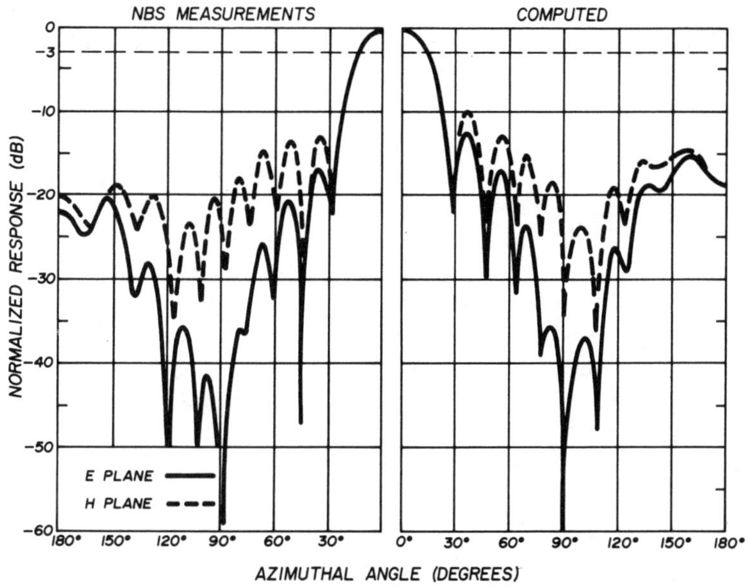

Fig. 1.12—Measured and computed *E*- and *H*-plane radiation patterns for a fifteen-element Yagi with a boom length of 4.2 wavelengths (from NBS Report 688 Fig. 19).

Table 1.5

Gain of seven different NBS-designed Yagi antennas determined by four different methods.

NBS Yagi	*calculated NBS measure-ments (dBi)*	*from half-power beamwidth (dBi)*	*pattern integration (dBi)*	*computer derived (dBi)*
2 el. (0.2 wavelength)	4.8	7.5	6.7	6.7
3 el. (0.4 wavelength)	9.3	10.0	9.6	9.2
5 el. (0.8 wavelength)	11.4	11.9	11.4	10.7
6 el. (1.2 wavelength)	12.4	13.9	12.6	11.8
12 el. (2.2 wavelength)	14.4	15.3	14.3	14.0
17 el. (3.2 wavelength)	15.6	16.6	15.5	15.2
15 el. (4.2 wavelength)	16.4	17.4	16.2	15.7

The gain of each of these Yagi designs can be obtained in several ways, and it is illuminating to compare all methods. Table 1.5 shows a comparison of three experimental methods all derived from the NBS data, and my theoretical calculations. Column 1 shows the NBS measured gain converted to isotropic gain (dBi). Column 2 shows the gain calculated from the measured half-power main beam angles by the usual formula, dBi $= 10 \log (41253/\theta_H\theta_E)$, where the H and E half-power angles are measured in degrees (Kraus, ref. 3, page 25). The third column is derived entirely from the experimental NBS patterns; in each case I calculated the directivity by appropriately summing all 10 degree intervals. This pattern averaging, if carefully done, should yield a reasonably reliable result, free from any systematic error (due to ground reflections, for example). The last column is the result of my theoretical calculations made on each design.

In Table 1.5, note that column 1 (NBS measurements) and column 4 (theory) are in agreement within Viezbicke's estimated measurement accuracy of about 0.5 dB with the exception of the value for the two-element Yagi (discussed earlier). Column 2 is derived from measured half-power angles; it is quite difficult to determine such angles with any great precision and, moreover, the method is nothing more than a crude approximation at best. Column 3 (derived from NBS experimental patterns) is in remarkable agreement with theory in all cases, especially considering that the summation in 10 degree intervals is really too coarse and that values in these intervals were "eyeballed" from the published NBS patterns.

Effect of Director Length

Fig. 7 in the NBS report shows an interesting experimental result: the selected designs are superior in gain to simplified Yagis (all directors of equal length) for booms longer than about one wavelength. Viezbicke did not give the lengths of the directors used for the simplified Yagis; I made calculations

Table 1.6
Measured and calculated gain of simplified Yagis with equal-length directors.

NBS Yagi	*NBS director length (wavelength)*	*measured gain (dBi)*	*computed gain (dBi)*
5 el. (0.8 wavelength)	0.4260	11.3	10.7
6 el. (1.2 wavelength)	0.4240	12.2	11.7
12 el. (2.2 wavelength)	0.4017	13.9	13.6
17 el. (3.2 wavelength)	0.3946	14.8	14.7
15 el. (4.2 wavelength)	0.4008	15.6	15.2

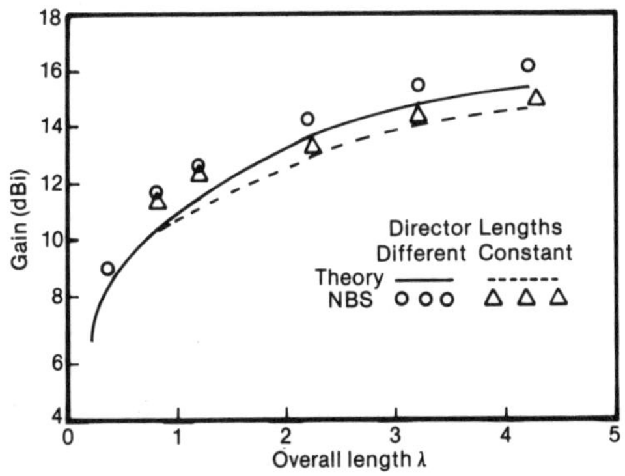

Fig. 1.13—Gain vs. boom length for equal-length and optimized-length directors.

using directors with lengths that are the average of those used in each of the five longer Yagis. His measurements of gain and my theoretical calculations are shown in Table 1.6. Again, these results are in satisfactory agreement.

Fig. 1.13 shows a graph of my theoretical results for both Tables 1.5 and 1.6, together with the published NBS experimental points. Note that the theoretical results support the idea that while the simplified (equal director length) Yagi is just as good as a more sophisticated design for booms shorter than one wavelength, a slight gain improvement in the gain of long Yagis is apparently possible by using directors of different lengths.

Viezbicke's Fig. 8, reproduced here as Fig. 1.14, shows the results of an interesting experiment in which gain for a given Yagi design was measured using a series of different director lengths, for each of eight element diameters. These measurements allow an interesting test of the theory of the reactance of cylindrical elements. In principle, a fiducial point on each gain curve, such as the length to produce peak gain, should correspond to the same element reactance for all eight cases. The reactance of the directors can be calculated from (1.7). From Fig. 1.14 it is somewhat difficult to estimate accurately the position of peak gain; therefore I instead estimated the (longer) director length which drops the gain to 9 dBd. Table 1.7 shows the results. Note that the computed reactance is essentially the same for all eight cases; the variances from the mean value of −60.8 ohms are not systematic nor are they

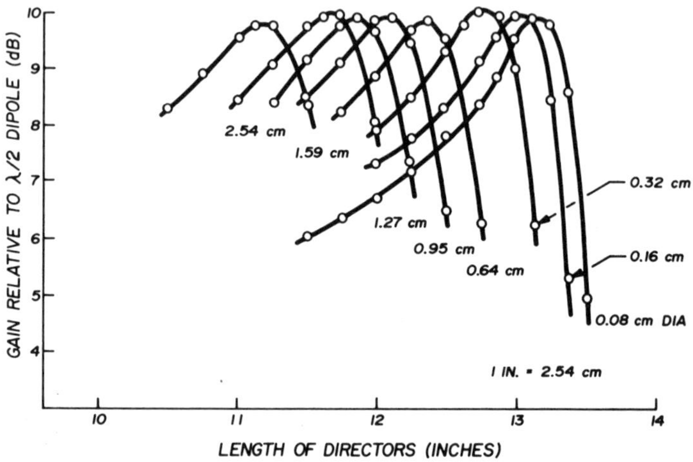

Fig. 1.14—Measured gain vs. director length at 400 MHz for a Yagi 1.25 wavelengths long with three directors spaced 0.35 wavelength.

Table 1.7

Element lengths (estimated from Fig. 1.14) which produce 9 dBd gain, their resonant lengths and their reactance. The average reactance is −60.8 ohms with standard deviation of 2.1 ohms.

element diameter (cm)	*element length (inches)*	*l (wavelength)*	*l_R (wavelength)*	*X_l (ohms)*
0.08	13.34	0.45171	0.48139	−63.6
0.16	13.22	0.44766	0.47896	−58.9
0.32	13.00	0.43988	0.47579	−58.3
0.64	12.58	0.42586	0.47147	−62.2
0.95	12.31	0.41699	0.46824	−62.3
1.27	12.09	0.40936	046535	−62.0
1.59	11.91	0.40329	0.46272	−60.8
2.54	11.43	0.38725	0.45565	−58.0

large. The standard deviation corresponds to an element length variation of less than 0.4 percent; such a variation could easily arise from length estimates in Fig. 1.14 or even in physically constructing the elements used in the experiment. Thus this experiment seems to give strong support for the theory of cylinder reactance.

Gain Variations

Perhaps one of the most interesting experimental results of the NBS work is the strange "oscillating" gain versus boom length characteristic shown in

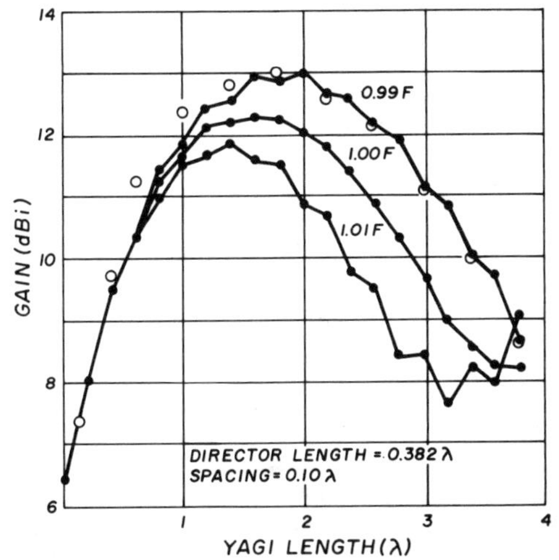

Fig. 1.15—Gain vs. boom length for small variations in normalized frequency (from NBS Report 688 Fig. 4).

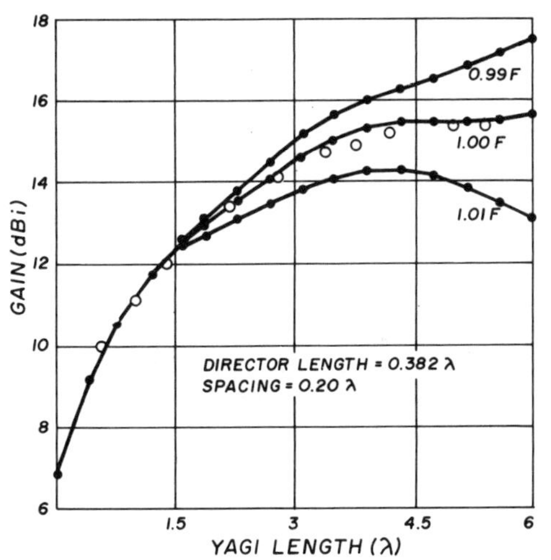

Fig. 1.16—Gain vs. boom length for small variations in normalized frequency (from NBS Report 688 Fig. 4).

Viezbicke's Figs. 4, 5, and 6. I attempted to calculate a number of these cases and although the oscillating gain phenomenon does show up, the detailed agreement between experiment and theory is not impressive. However, I found that the exact behavior of these long and heavily (director) loaded arrays is critically dependent on frequency and/or exact director lengths. Accordingly, I ran calculations for several frequencies around the central frequency (normalized to unity). The results are shown in Figs. 1.15 through 1.19, where the theoretical calculated points for each frequency are connected by line segments for clarity. Superimposed on the plots are Viezbicke's experimental points.

The theoretical results generally show that the accurately computed points do not really lie on a smooth curve; this is the reason connecting lines are used. Theory does give the oscillating gain phenomenon, but it also shows that the details depend sharply on the exact transmitter frequency. Pretty good agreement can be obtained between experiment and theory of it is assumed that slightly different frequencies ($\approx$ 1 to 3 percent) were used from experiment to experiment. It is perhaps significant that the most likely frequency does not appear to be systematically high or systematically low.

The theoretical results are interesting from three points of view. First, for a given frequency the calculated points do not appear to lie on a smooth curve; it is likely that the gain varies somewhat with side- and back-lobe structure. The calculated *F/B* ratio varies greatly from point to point, and it would seem reasonable that there would be some reaction on forward gain. Second, for long Yagis the results vary significantly with the exact frequency used. Third, detailed calculations show that beyond the end of the first "oscillating" gain cycle, Yagi gain is not maximum in the plane of the Yagi; instead, the front beam appears to develop a central dimple causing maximum gain to occur at an appreciable elevation angle. (The plots in Figs. 1.15 through 1.19, however, show only the gain at zero elevation angle because this is what was presumably done in the NBS experiments.)

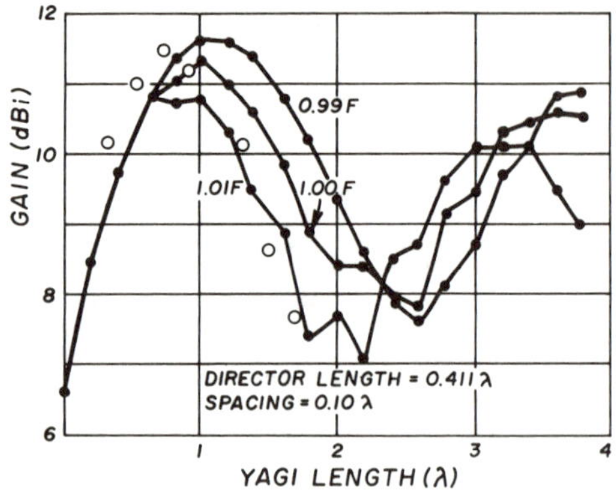

Fig. 1.17—Gain vs. boom length for small variations in normalized frequency (from NBS Report 688 Fig. 5).

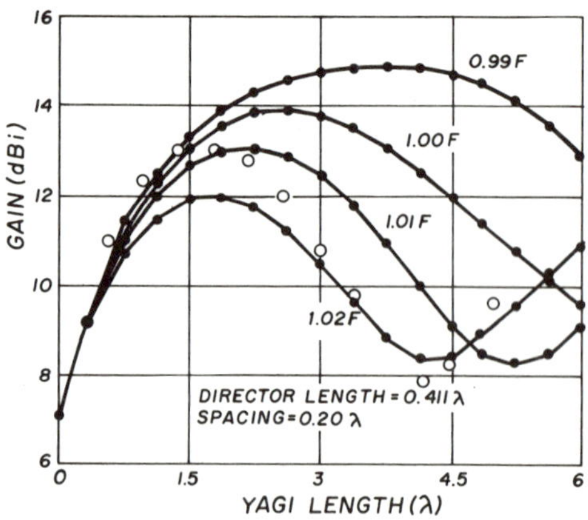

Fig. 1.18—Gain vs. boom length for small variations in normalized frequency (from NBS Report 688 Fig. 5).

Note that all three of these behavioral aspects of long Yagis appeared from theoretical computations; they were not revealed by the NBS experiments. This illustrates the point that an extensive variety of Yagi configurations can be explored much faster theoretically, and with much more relative accuracy, than can be accomplished experimentally. As a result, new concepts and understanding emerge.

Comparison Summary

Let me summarize the overall comparison of the experimental results in the NBS report with the theoretical results. The comparisons show total agreement on Yagi gain for an astonishing range of models with the single exception of the direct measurements on a two-element beam. A gain figure derived from the measurement of the pattern of the two-element beam, however, does agree with theory; it is my belief that an error was made by the NBS group in their measurements.

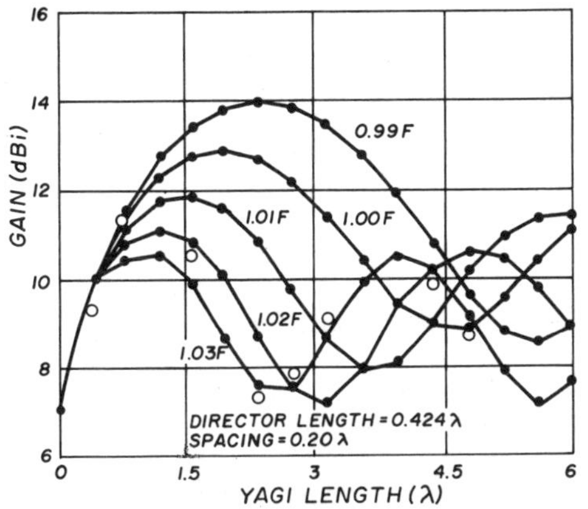

Fig. 1.19—Gain vs. boom length for small variations in normalized frequency (from NBS Report 688 Fig. 6).

The comparisons also show excellent agreement for Yagi patterns, again over a wide range of models. This should be a very sensitive test of the accuracy with which mutual and self-impedances are represented, as well as the resonant frequency calculations of cylindrical elements. This latter point is also strongly supported by the consistent values of parasitic element reactance for elements whose diameters vary over a range of greater than 30 to one.

Finally, the strange "oscillating" gain phenomenon observed by Viezbicke can be reproduced theoretically. Agreement is only qualitative if the NBS group used an accurate 400.0 MHz frequency in their tests; however, the comparisons suggest that they may have actually used a nominal "400 MHz," with frequencies within a range of ± 3 percent. In this event, the comparisons indicate a potentially remarkable agreement in quantitative aspects as well.

Since these comparisons include very long Yagi models and correspondingly large numbers of parasitic elements, it seems certain that the computational methodology I have outlined should be generally trustworthy. This should be especially true for shorter antennas (with fewer elements) with which I am primarily concerned throughout this book. Especially noteworthy is the fact that all computed results contain *no adjustable constants or parameters*; they are all derived from basic physical principles using adequately accurate mathematical approximations.

NOTES:

[1]Yagi, H. and Uda, S., *Proceedings of the Imperial Academy*, February, 1926. Also Uda, S., *Journal of the Institute of Electrical Engineers of Japan*, Volume 47-48, 1927-1928.

[2]Uda, S. and Mushiake, Y., *Yagi-Uda Antenna*, Research Institute of Electrical Communication, Tohoku University, Sendai, Japan: Sasaki Ltd., 1954.

[3]Kraus, John D., *Antennas*, New York: McGraw-Hill, 1950.

[4]King, Ronald W. P., *Theory of Linear Antennas*, Cambridge, Massachusetts: Harvard University Press, 1956.

[5]Walkinshaw, W., "Treatment of Short Yagi Aerials," *Journal of the IEE (London)*, Volume 93, Part 3A, Number 3, 1946.

[6]Ehrenspeck, H. and Poehler, H., "A New Method of Obtaining Maximum Gain from Yagi Antennas," *IRE Transactions on Antennas and Propagation*, October, 1959.

[7]Lindsay, J., "Quads and Yagis," *QST*, May, 1968.

[8]Greenblum, C., "Notes on the Development of Yagi Antennas," *QST*, Part 1, August, 1956; Part 2, September, 1956.

[9]Viezbicke, P., "Yagi Antenna Design," *NBS Technical Note 688*, U.S. Department of Commerce, Washington, D.C., December, 1976.

[10]"IEEE Standard Test Procedures for Antennas, IEEE Std 149-1979," The Institute of Electrical and Electronic Engineers, Inc., December 1979, page 94.

[11]Hallen, E., "Theoretical Investigations into the Transmitting and Receiving Qualities of Antennae," *Nova Acta Uppsala (Sweden)*, Series IV, Volume 11, Number 4, 1938.

[12]Any improvement in these approximations for self- and mutual impedances will require a great amount of theoretical work through a rigorous examination of the boundary value problem with attention to (1) current distributions along driven and parasitic elements (they are somewhat different in principle), and (2) complete numerical solutions to both real and imaginary components of element self and mutual impedance. It will be necessary to distinguish mutual coupling coefficients between elements of different function; i.e., driver-to-driver, driver-to-parasite, and parasite-to-parasite. In principle, all coefficients will depend on each affected element length and radius.

[13]Carter, P., "Circuit Relations in Radiating Systems and Application to Antenna Problems," *Proceedings of the IRE*, 20, June, 1932.

[14]Mutual impedance can also be calculated directly from the exponential integral equations—see Kraus, ref. 3, page 265.

CHAPTER 2

SIMPLE YAGI ANTENNAS

A Yagi antenna can be characterized by one or more driver elements and a number of parasite elements, all supported on a boom. For each element we must specify x and y coordinates, a length, and a radius, all measured in terms of wavelength at the design frequency. In the case of each driver we must also specify the excitation potential or current.

It is instantly apparent that with all these variables an exhaustive investigation into all possible configurations is impractical. Instead I shall begin with an initial consideration of simplistic Yagi antennas and will subsequently discuss a variety of departures from the basic design. This class of Yagi antennas has (1) a single driven element; (2) one reflector; (3) zero or more directors with identical lengths and diameters; and (4) all elements uniformly spaced along the boom. Additionally, the antenna will be in free space.

These restrictions may seem at first sight to be quite severe, but I hasten to remark that free space performance relates to actual performance over ground and that simplistic Yagi antennas, as defined above, can, in many instances, provide performance levels fully as good as those from more sophisticated designs. Furthermore, we can learn a great deal about Yagi antenna performance from studying these simple designs and, as we shall see, will develop useful conceptual ideas about Yagi behavior and ideas for "best" design.

Throughout this investigation of simplistic antennas I will choose element dimensions (radii) characteristic of "normal" 14 MHz construction— $a = 0.000526\lambda$ corresponding to 7/8 inches diameter. The results can be translated to any other element dimension by proper scaling calculations; scaling rules are given in Chapter 7.

Two-Element Beams

I shall begin with a two-element Yagi beam involving one parasite which can act either as a reflector (for frequencies above its resonance) or as a director. For such a beam there are only two fundamental variables, the physical separation of the two elements along the boom and the physical length of the parasite. The exact length of the driven element is of little consequence as far as gain and pattern are concerned; it does, however, affect driving-point impedance (especially reactance) which is considered later. Since we are interested in a frequency-swept plot of the gain and *F/B* properties, the physical length of the parasite can be fixed. As the frequency increases from well below to well above the parasite free space resonant frequency we can observe the properties of the beam first where the parasite behaves as a director and then as a reflector.

The computation methodology is that given explicitly in Chapter 1. In all cases of the two-element beam the element lengths are 0.48167λ for both parasite and driver; this makes each element's normalized free space resonant frequency F_R equal to unity. The normalized frequency, *F*, is varied in steps from about 0.90 to 1.10; these steps were made sufficiently small to show the detailed behavior of the beam. Element separation, *s*, measured in wavelengths at the design frequency, is varied from 0.025λ to 0.5λ, again in steps sufficiently small to bring out essential behavior.

Fig. 2.1 shows the frequency-swept gain and front-to-back ratio of the two-element Yagi antenna for several element spacings, where the gain is posi-

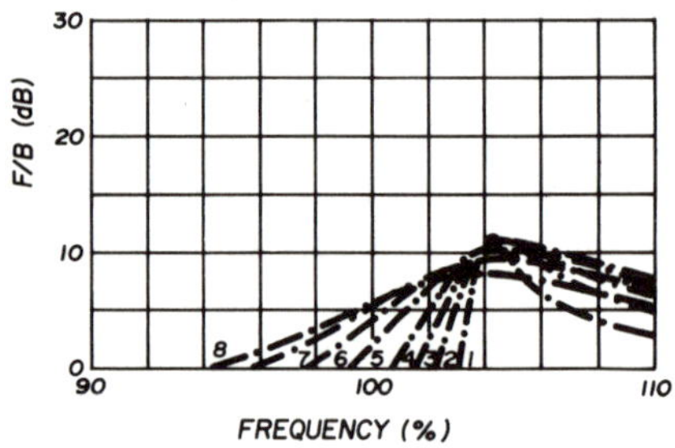

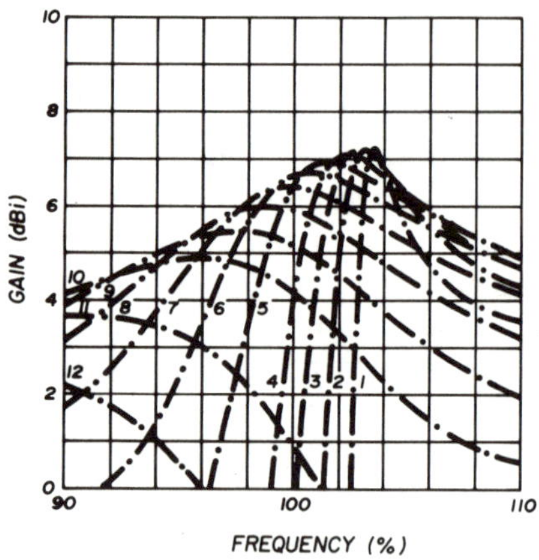

Curve	s	Curve	s	Curve	s
1	0.025	5	0.150	9	0.350
2	0.050	6	0.200	10	0.400
3	0.075	7	0.250	11	0.500
4	0.100	8	0.300	12	0.600

Fig. 2.1—Frequency-swept gain and front-to-back ratio of two-element Yagi antennas with different element spacings, where the parasitic element acts as a reflector.

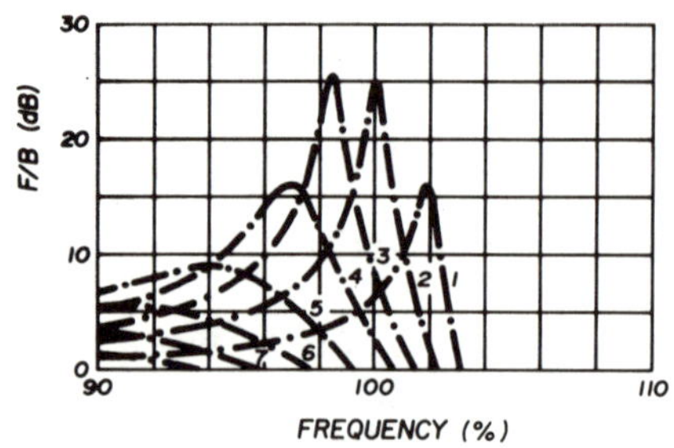

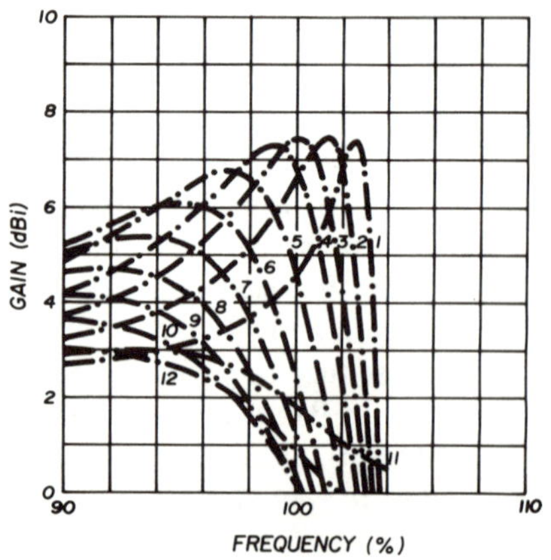

Curve	s	Curve	s	Curve	s
1	0.025	5	0.150	9	0.350
2	0.050	6	0.200	10	0.400
3	0.075	7	0.250	11	0.500
4	0.100	8	0.300	12	0.600

Fig. 2.2—Frequency-swept gain and front-to-back ratio of two-element Yagi antennas with different element spacings, where the parasitic element acts as a director.

tive in the direction of parasite towards driver (that is, where the parasite acts like a reflector). Each curve represents a particular spacing as keyed in the legend. Fig. 2.2, in the same format, shows the gain and *F/B* where the parasite acts like a director, i.e., where the gain is positive in the direction of driver towards parasite.

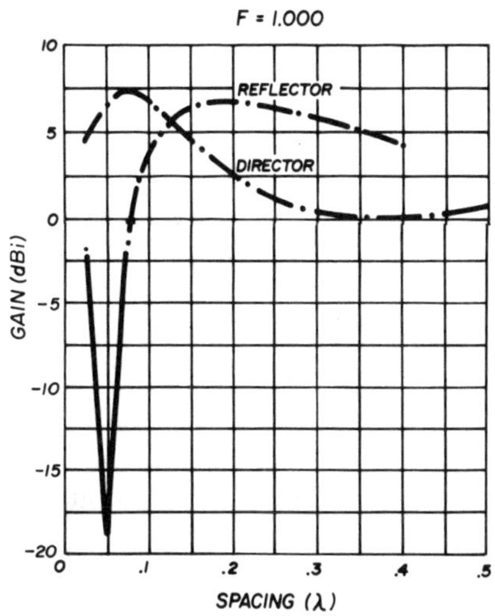

Fig. 2.3—Gain of two element Yagis at the design frequency (F = 1) as a function of element spacing.

Examination of these performance plots, together with additional information on computed driver input impedance, reveals a number of interesting facets of the behavior of two-element Yagi beams. Fig. 2.3 shows a plot of the gain at $F = 1$ as a function of element spacing. This is similar to but not identical with a plot shown on page 147 of *The ARRL Antenna Book*.[1] It is possible that the differences, particularly at small spacings, are due to greater precision in the new calculations; in any case, I believe the implications of this plot are somewhat misleading.

One can easily see that the maximum gains obtainable at the "best" frequency(s) in Figs. 2.1 and 2.2 look somewhat different. These are shown in Figs. 2.4 and 2.5, where for reference the curve for $F = 1.0$ is also shown. Note that the obtainable gain does not depend greatly on whether the parasite is a reflector or a director as implied in Fig. 2.3; moreover, the largest gain is obtained at very small spacings. This is a result which is not intuitive. Figs. 2.4 and 2.5 can be compared with the early analysis by Brown, and there appears to be good agreement.[2] However, the result shown on page 146 of *The ARRL Antenna Book* which cites Brown's analysis is somewhat different. The fall off in maximum gain for low spacings shown by the ARRL

[1]This and other footnotes are given at the end of this chapter.

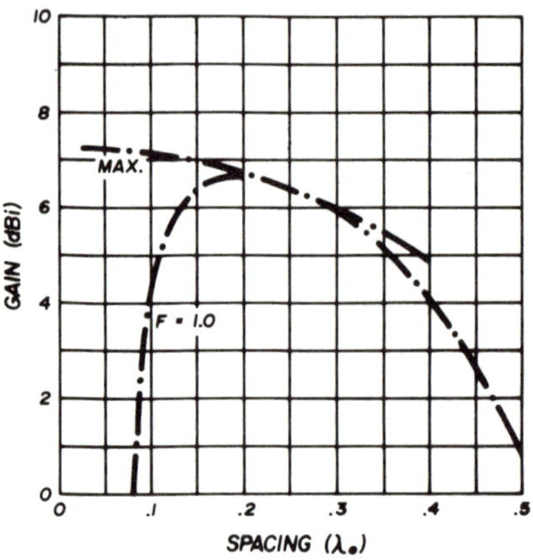

Fig. 2.4—Gain of two-element Yagis vs. element spacing; parasite a reflector.

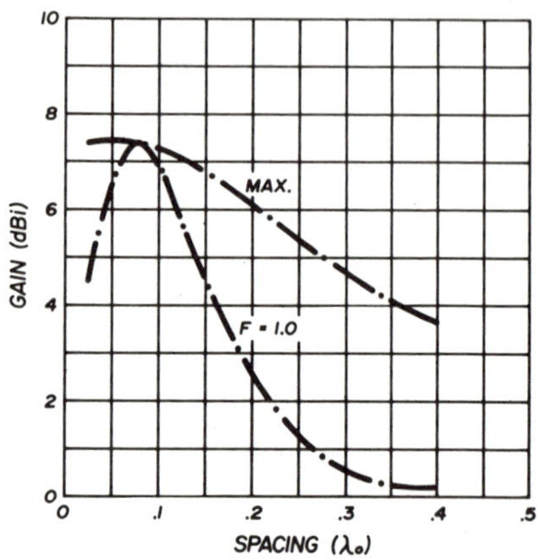

Fig. 2.5—Gain of two-element Yagis vs. element spacing; parasite a director.

reference does not appear to agree with Brown nor does it substantiate the calculations I have made (Figs. 2.4 and 2.5).

If one examines the maximum gain shown in Figs. 2.1 and 2.2 for best frequency and corresponding (driver) driving-point impedance, one obtains the values shown in Table 2.1. A plot of *R* is shown in Fig. 2.6 which can be compared with a similar diagram shown on page 147 of *The ARRL Antenna Book*; except for low values of spacing, agreement is fairly satisfactory.

One can see from Table 2.1 and Fig. 2.6 that element spacing affects

Table 2.1

Maximum gain and feed-point impedance of a two-element Yagi for various element spacings.

	parasite a reflector					*parasite a director*				
s (λ)	max gain (dBi)	at (freq)	R (Ω)	X (Ω)	Q	max gain (dBi)	at (freq)	R (Ω)	X (Ω)	Q
0.025	7.24	1.036	1.1	11.2	925	7.41	1.026	1.1	−10.0	1003
0.050	7.22	1.032	4.2	20.4	237	7.46	1.014	4.1	−17.6	266
0.075	7.16	1.030	9.7	31.8	99	7.42	1.000	9.8	−29.9	105
0.100	7.12	1.025	15.6	35.8	60	7.30	0.990	15.3	−34.1	63
0.150	6.96	1.015	29.6	38.6	29	6.80	0.970	27.5	−44.2	31
0.200	6.72	1.005	44.6	34.7	17	6.12	0.955	38.0	−49.4	21
0.250	6.41	1.000	61.4	31.6	11	5.38	0.940	48.2	−59.4	15
0.300	6.00	0.990	72.7	16.9	8	4.72	0.920	57.2	−80.5	12
0.350	5.49	0.980	80.5	−0.4	7	4.16	0.900	64.2	−104.2	11
0.400	4.91	0.960	82.0	−29.2	7	3.69	0.900	70.7	−103.3	9

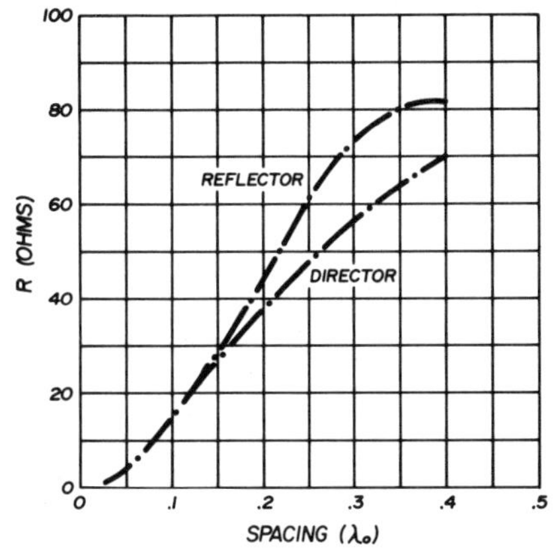

Fig. 2.6—Driving-point resistance of two-element Yagis vs. element spacing.

driving-point resistance, and therefore *Q*, over a very large range. This factor, as well as the gain curves shown in Figs. 2.1 and 2.2 set a practical limit to the achievable gain over a desirable bandwidth, for example, perhaps four percent in *F*. Moreover, the higher values of (radiation) loaded *Q* cause larger circuit resistive losses and therefore lower the antenna efficiency. Thus, in practice, really short booms are not desirable; one must choose between efficiency and bandwidth on the one hand, and gain and *F/B* ratio on the other.

Long booms, however, also appear undesirable because gain really falls off (primarily due to reduced excitation of the parasite). Furthermore, for booms longer than 0.3λ a new phenomenon can be seen from a detailed computational analysis (not shown here). The front lobe of radiation begins to "dimple" in the forward direction, resulting in a pattern where the gain maximum occurs at an elevation angle other than zero with respect to the boom direction. (The gain shown in Figs. 2.1 and 2.2, however, is just the directivity in the direction of the boom.) This pattern feature was predicted by Brown (ref. 2) and shown in Kraus, page 294.[3]

Note that the two-element Yagi gives respectable performance in gain for a wide range of element separations. However, the *F/B* figures are not especially impressive; moreover best *F/B* does not occur at the same frequency as best gain. Thus, in designing a two-element Yagi beam a practical compromise is necessary. If you wish to obtain good gain with at least a fair *F/B* ratio over a bandwidth of say four percent, you can determine by inspection of Figs. 2.1 and 2.2 and Table 2.1 that a two-element beam should have a boom length of perhaps 0.15 wavelengths. For such a boom the gain is essentially independent of whether the parasite is a reflector or a director; moreover, the *F/B* is about equivalent for either situation.

To move the peak of a gain curve in Figs. 2.1 and 2.2 to the design frequency, the parasite length is adjusted commensurately; to reduce design frequency driver reactance, the driver length is adjusted. These character-

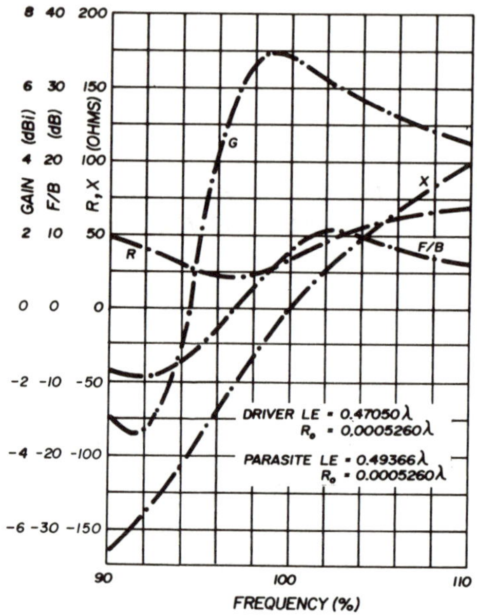

Fig. 2.7—Gain, front-to-back ratio and feed-point impedance as a function of frequency for a two-element Yagi with 0.15 wavelength spacing (parasite a reflector).

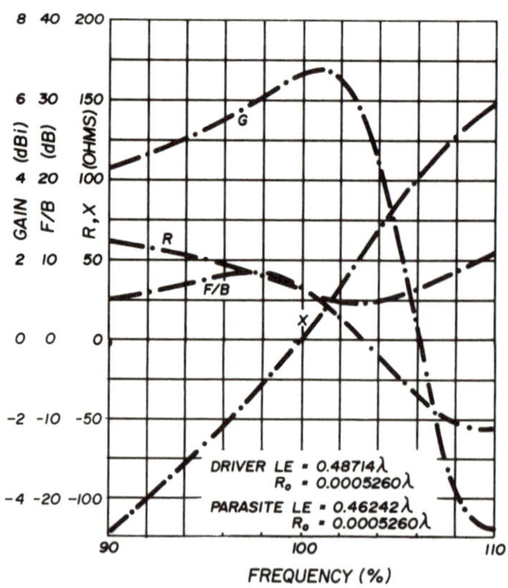

Fig. 2.8—Gain, front-to-back ratio and feed-point impedance as a function of frequency for a two-element Yagi with 0.15 wavelength spacing (parasite a director).

istics of two-element beams are shown in Figs. 2.7 and 2.8 with a frequency-swept plot of each design. Note that each of these figures shows gain, *F/B*, *R* and *X* of the driver. They illustrate the kind of design compromises which must be made. They also show the frequency-swept behavior of the main performance parameters.

The "best" design frequency is a matter of choice and is a compromise between gain and F/B ratio; it is adjusted by the length of the parasite. The driving-point resistance and reactance vary significantly with frequency. Note that you cannot generally specify resistance except at a single frequency; also note that feed-point reactance is not a linear function of frequency. This is caused by the combined effect of self- and mutual impedance of the elements.

One can adjust the frequency of the zero reactance point by adjusting the length of the driver; this has been done only approximately in Figs. 2.7 and 2.8. However, note that the adjusted driver lengths are quite different than the length for a single isolated resonant dipole in free space. These differences are again caused by mutual reactance coupled into the driver by the parasite.

More Than Two Elements

I now turn to an analysis of simplistic Yagi antennas having more than one parasite. In all cases there is only one reflector, all directors have identical lengths, and all elements are uniformly spaced along the boom. I shall display for clarity only the essential frequency-swept gain and frequency-swept F/B behaviors; the driving-point impedances, all of which were computed, are of secondary interest at this point. For these Yagi antennas there are a number of parameters which should be systematically explored. Table 2.2 shows these parameters and the range over which each has been varied. To display results in a consistent way I have chosen for each frequency-swept plot a fixed number of elements and a fixed overall boom length, l_B, measured in wavelengths at the design frequency. On each plot there are six numbered curves; each number designates a particular parasite "tuning" combination; these combinations are shown in Table 2.3. The lengths and free space resonant frequencies of parasites are shown for each numbered combination.

Table 2.2

Range of Yagi parameter variations.

parameter	*range*
Number of elements	3,4,5,6,7
Boom length	up to 1.5λ
Reflector resonant frequency	0.98 to 0.93
Director resonant frequency	1.02 to 1.07

Table 2.3

List of parasitic lengths and resonant frequencies for the six numbered curves on Figs. 2.9 to 2.11.

	reflector		*director*	
curve	*length* (λ)	*resonance*	*length* (λ)	*resonance*
1	0.49150	0.98	0.47223	1.02
2	0.49657	0.97	0.46764	1.03
3	0.50174	0.96	0.46314	1.04
4	0.50702	0.95	0.45873	1.05
5	0.51241	0.94	0.45441	1.06
6	0.51792	0.93	0.45016	1.07

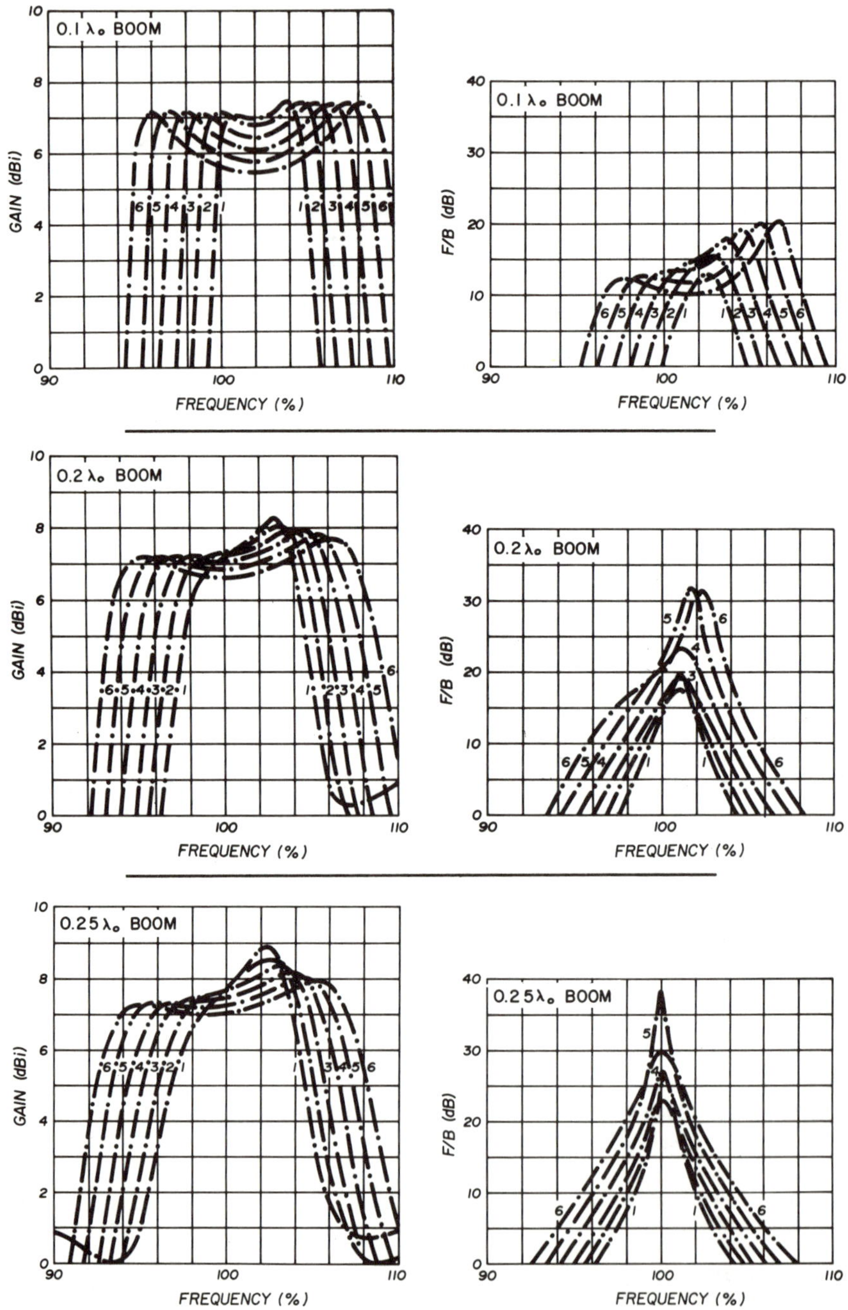

Fig. 2.9—Gain and front-to-back ratio for three-element Yagi beams with different boom lengths and parasitic element lengths (See Table 2.3).

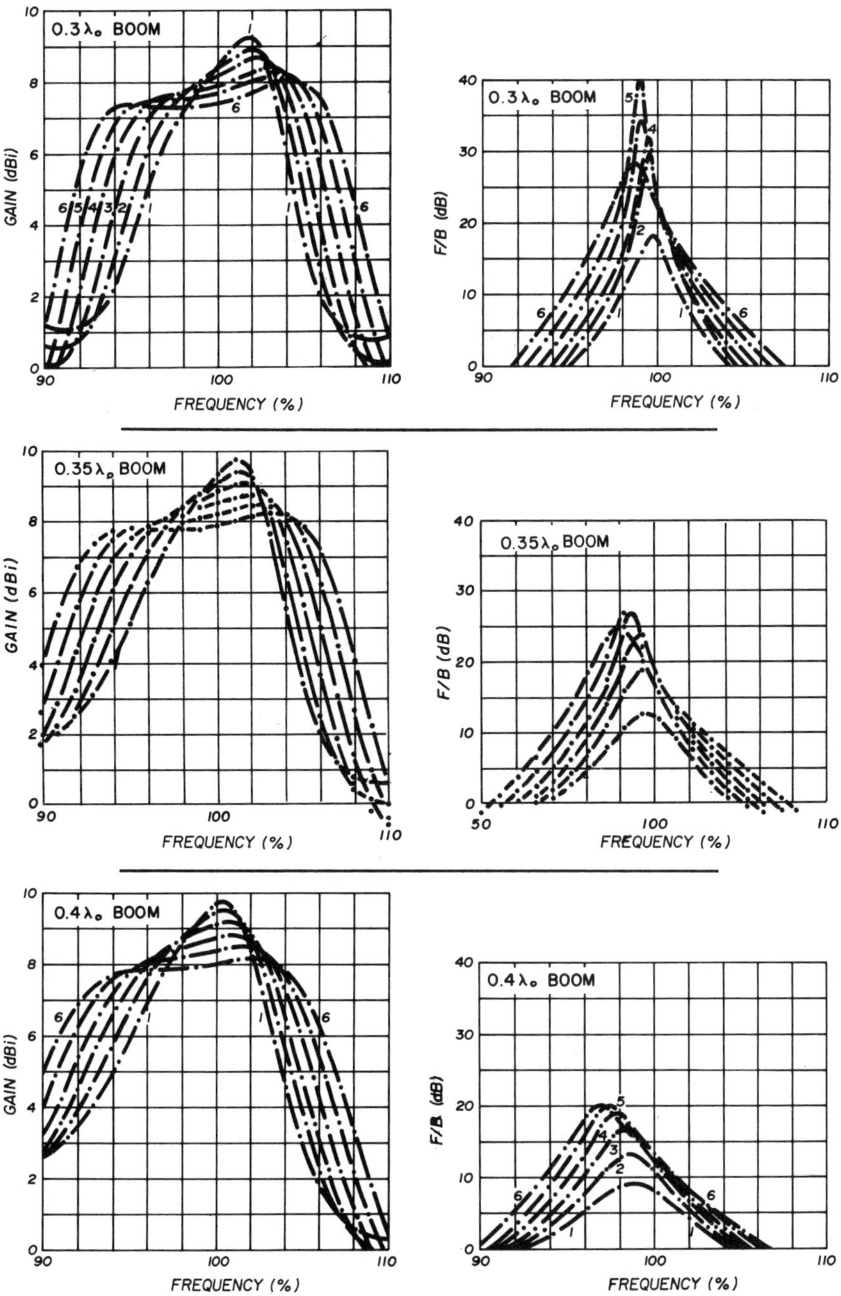

Fig. 2.9 (continued)—Gain and front-to-back ratio for three-element Yagi beams with different boom lengths and parasitic element lengths (See Table 2.3).

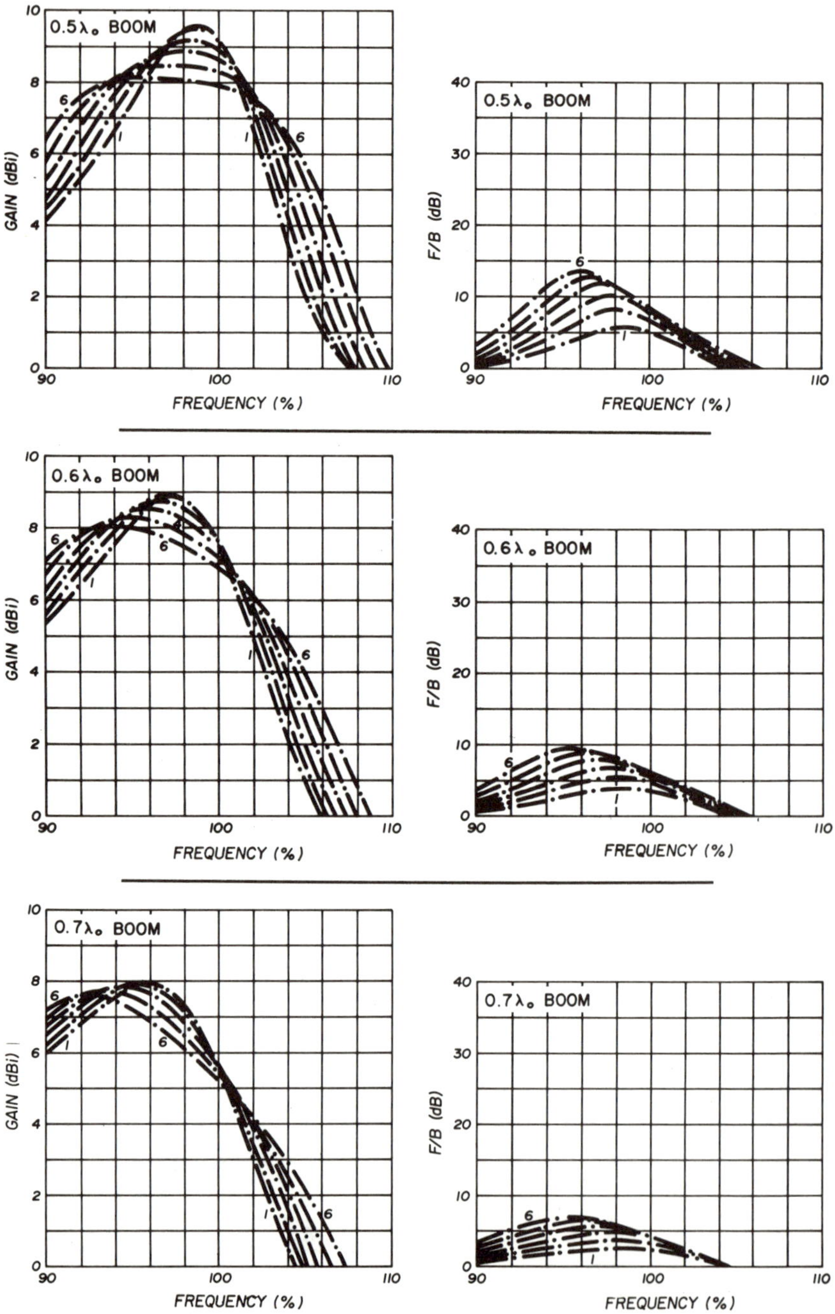

Fig. 2.9 (continued)—Gain and front-to-back ratio for three-element Yagi beams with different boom lengths and parasitic element lengths (See Table 2.3).

The curves of Fig. 2.9 show the results for three-element beams as boom lengths are varied from 0.100 to 0.700 wavelengths. It is apparent from an inspection of these plots that the performance is superior to that of the two-element beams; this is especially true in the *F/B* ratio. As one increases boom length the maximum gain increases (unlike that for two-element beams); the *F/B* increases spectacularly, then decreases again. For this class of Yagi antennas there seems to be a best boom length; we shall see this kind of result for all of the simplistic Yagis and the physical explanation will soon be apparent.

Note that the chief parameter controlling the bandwidth over which gain remains high is the (resonant) frequency separation of reflector and director; this observation will also prove to be generally true for all simplistic Yagi antennas. The bandwidth of the *F/B* performance (when the *F/B* is very high) is small; this is due to the critical nature of low back radiation. Back radiation is very low only when there is vectorial cancellation of field in the back direction; this comes about only where element complex currents are accidentally favorable for such cancellation. When this happens, very small changes in those currents, e.g. shifting frequency slightly, will destroy the favorable vectorial cancellation. This general result is inherent in all Yagi antennas; if the *F/B* is exceptionally high, it will be so only over a very narrow frequency band.

Similar results for four-element simplistic beams are shown in Fig. 2.10, and for six-element simplistic beams in Fig. 2.11. A quick inspection of these results shows the increasing complexity of the frequency-swept plots with number of elements; this is caused primarily by the larger number of resonances in the system. The gain "cutoff" at high frequencies is also increasingly abrupt due no doubt to the large number of directors (which shift over to "reflectors" at high frequencies). *F/B* curves become very complex. Moreover, really high values of *F/B* (greater than 30 dB) are quite rare; it is very unusual to find combinations where vectorial cancellation in the rear direction is nearly complete.

The total range of results shows a number of characteristics of interest. First, the bandwidth over which gain is high is determined primarily by the frequency spread between the reflector and the director(s). Second, the shape of the gain curve is generally not flat in the region of interest; indeed it may be sloped and/or humped or dished. Usually the slope favors the higher frequencies. Third, the shape of the gain curve is more complex where the number of elements is large, and the shape of the *F/B* ratio varies enormously—much more than the shape of the associated gain curve. It may show more than one peak; moreover, the peak structure shifts very rapidly with boom length. The height of the peak does not necessarily seem to vary monotonically with the frequency separation of reflector and director(s). Very high *F/B* values (greater than 30 dB) are quite rare and when present are invariably very narrow-banded. In addition, the frequency bandwidth of the *F/B* parameter is undefinable because of the extreme variation in shape.

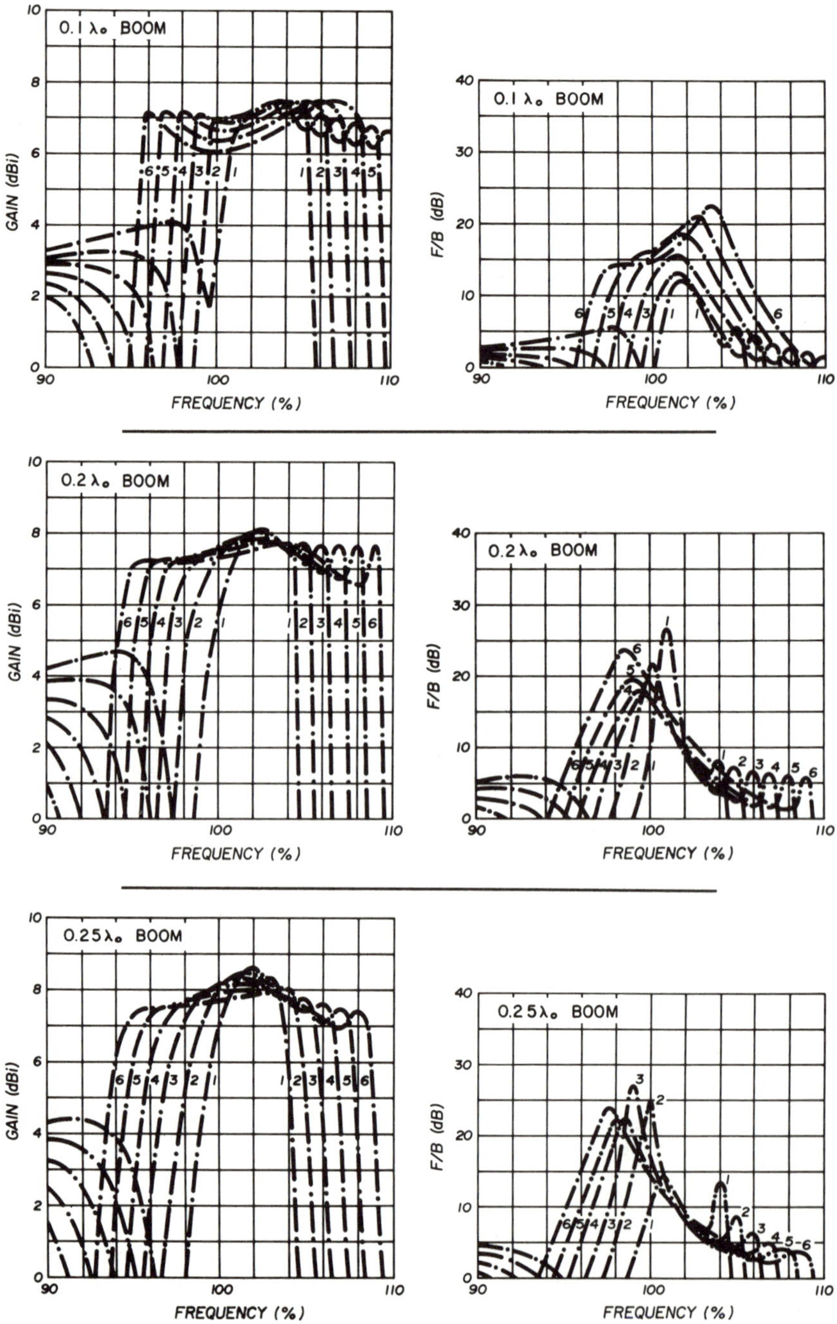

Fig. 2.10—Gain and front-to-back ratio for four-element Yagi beams with different boom lengths and parasitic element lengths (See Table 2.3).

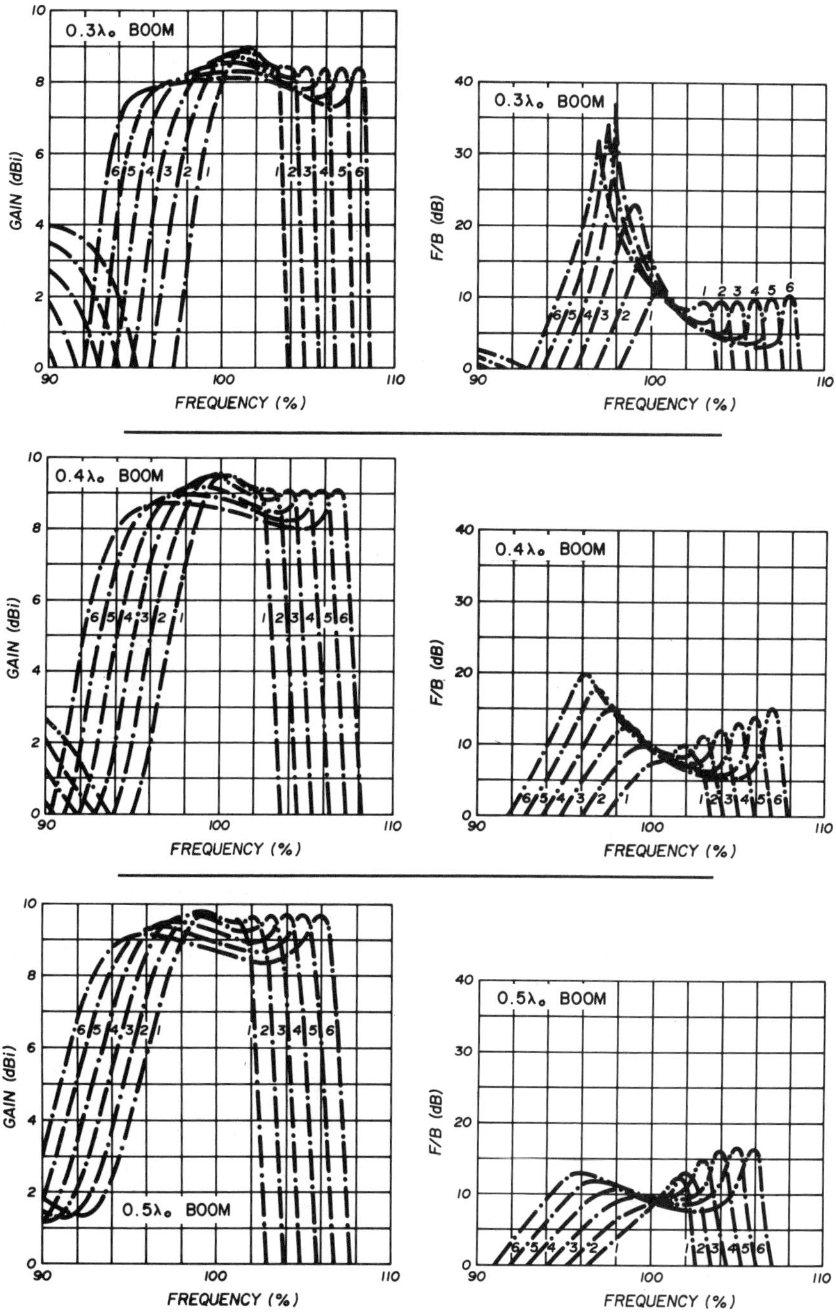

Fig. 2.10 (continued)—Gain and front-to-back ratio for four-element Yagi beams with different boom lengths and parasitic element lengths (See Table 2.3).

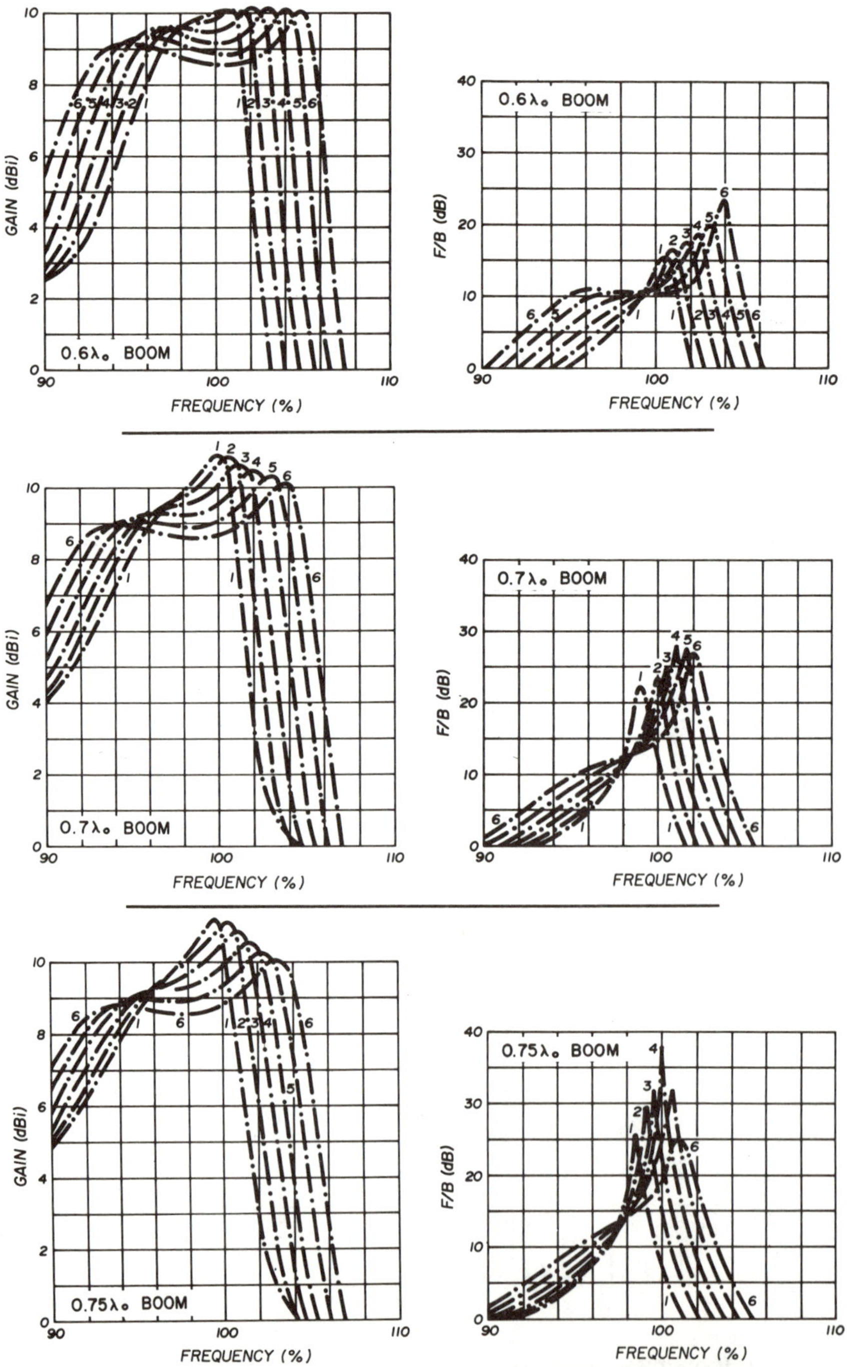

Fig. 2.10 (continued)—Gain and front-to-back ratio for four-element Yagi beams with different boom lengths and parasitic element lengths (See Table 2.3).

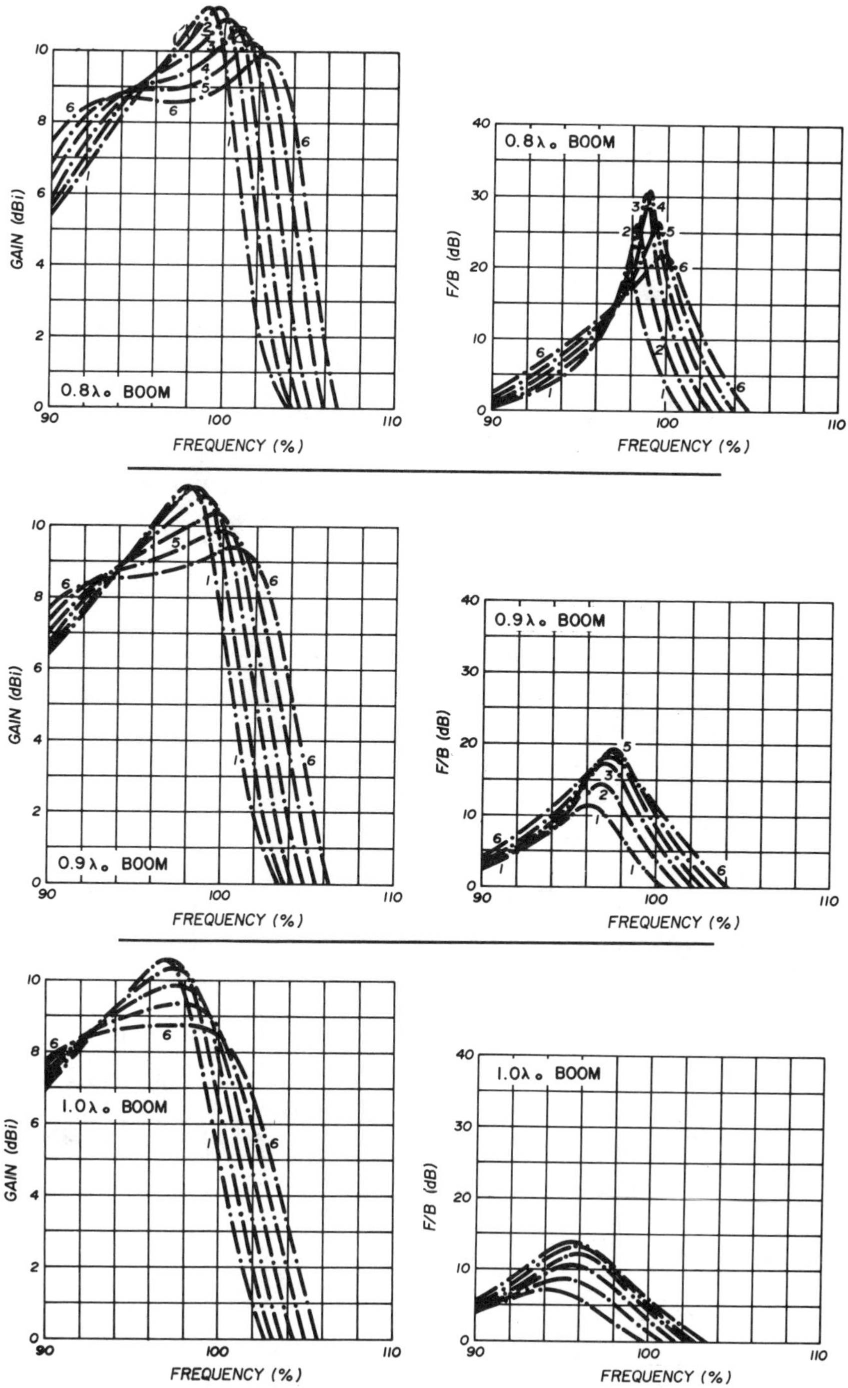

Fig. 2.10 (continued)—Gain and front-to-back ratio for four-element Yagi beams with different boom lengths and parasitic element lengths (See Table 2.3).

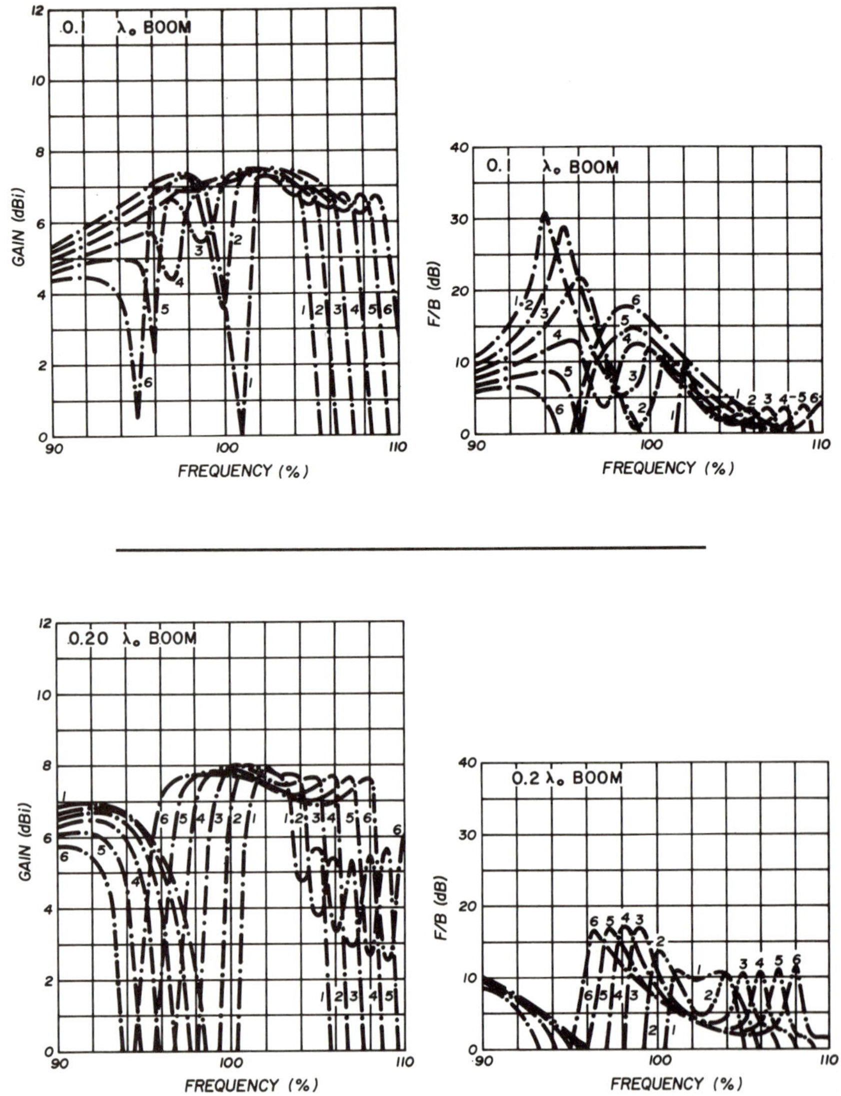

Fig. 2.11—Gain and front-to-back ratio for six-element Yagi beams with different boom lengths and parasitic element lengths (See Table 2.3).

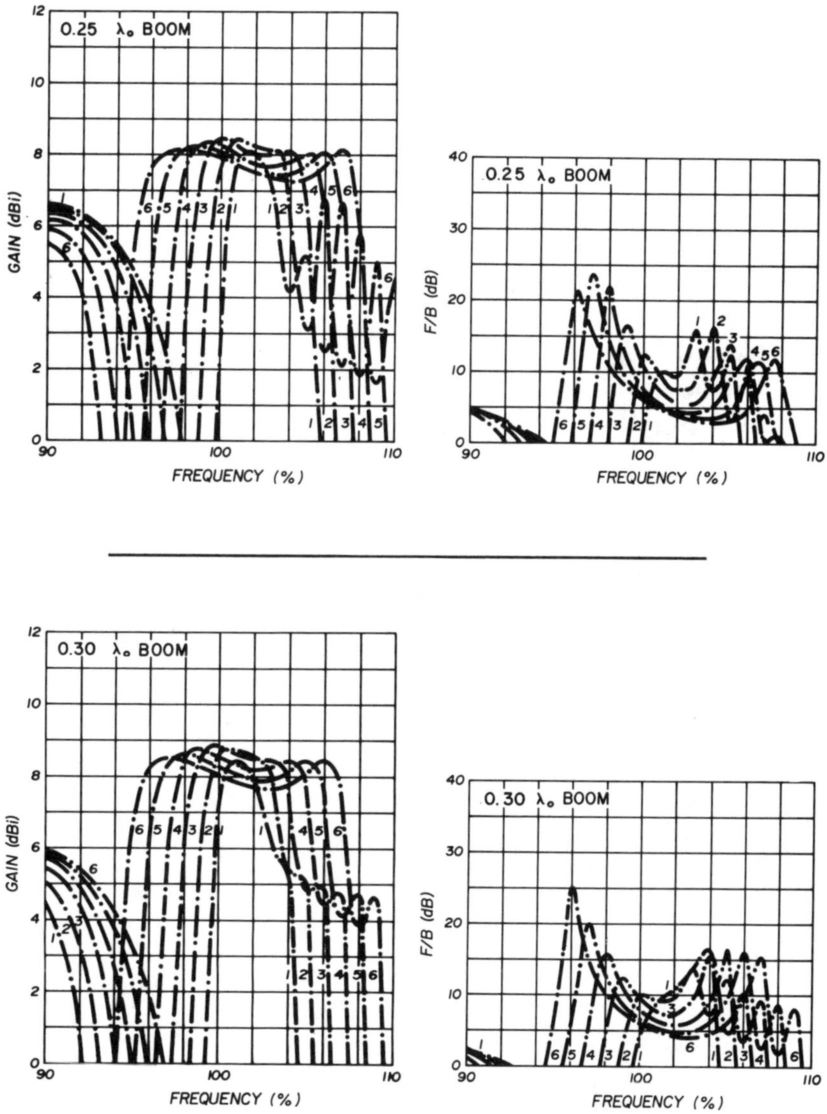

Fig. 2.11 (continued)—Gain and front-to-back ratio for six-element Yagi beams with different boom lengths and parasitic element lengths (See Table 2.3).

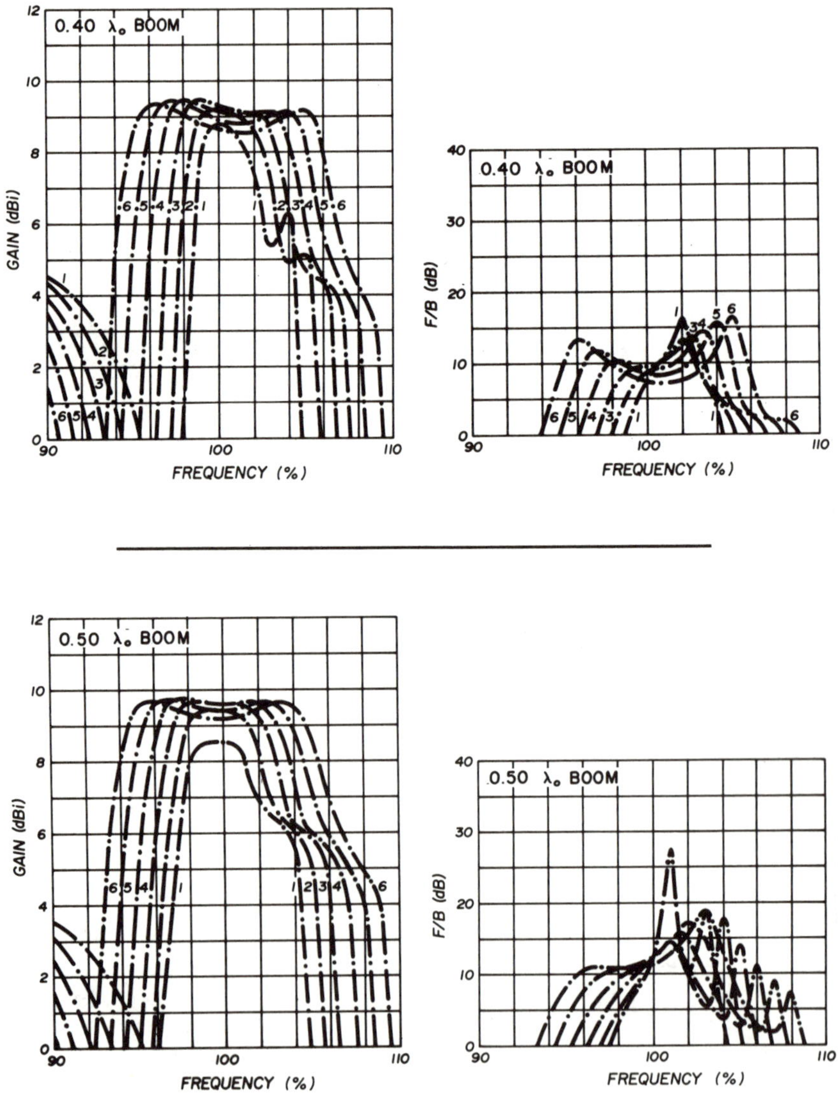

Fig. 2.11 (continued)—Gain and front-to-back ratio for six-element Yagi beams with different boom lengths and parasitic element lengths (See Table 2.3).

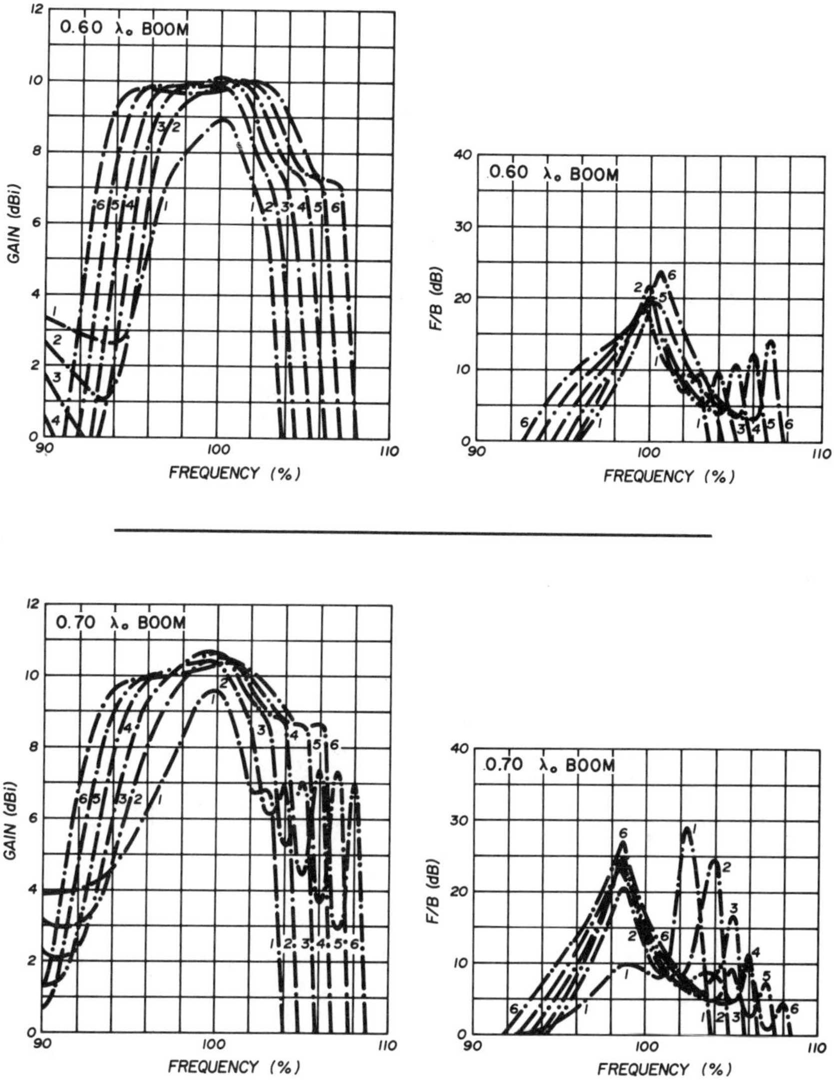

Fig. 2.11 (continued)—Gain and front-to-back ratio for six-element Yagi beams with different boom lengths and parasitic element lengths (See Table 2.3).

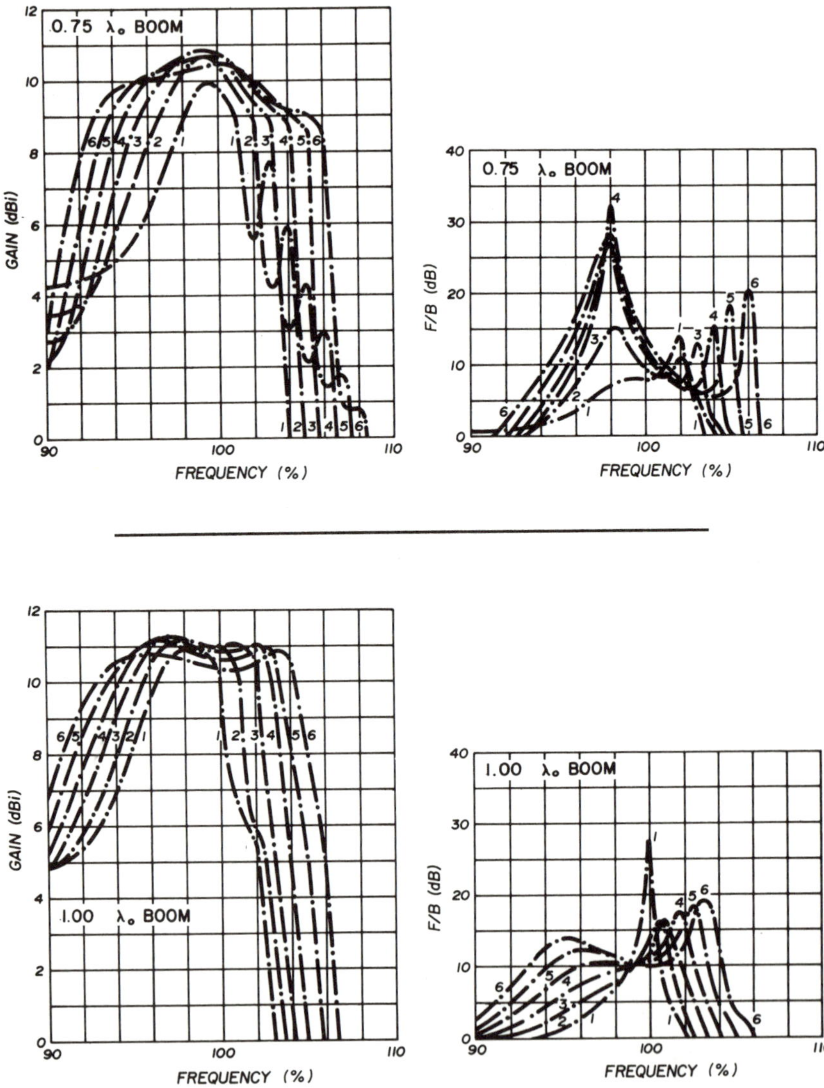

Fig. 2.11 (continued)—Gain and front-to-back ratio for six-element Yagi beams with different boom lengths and parasitic element lengths (See Table 2.3).

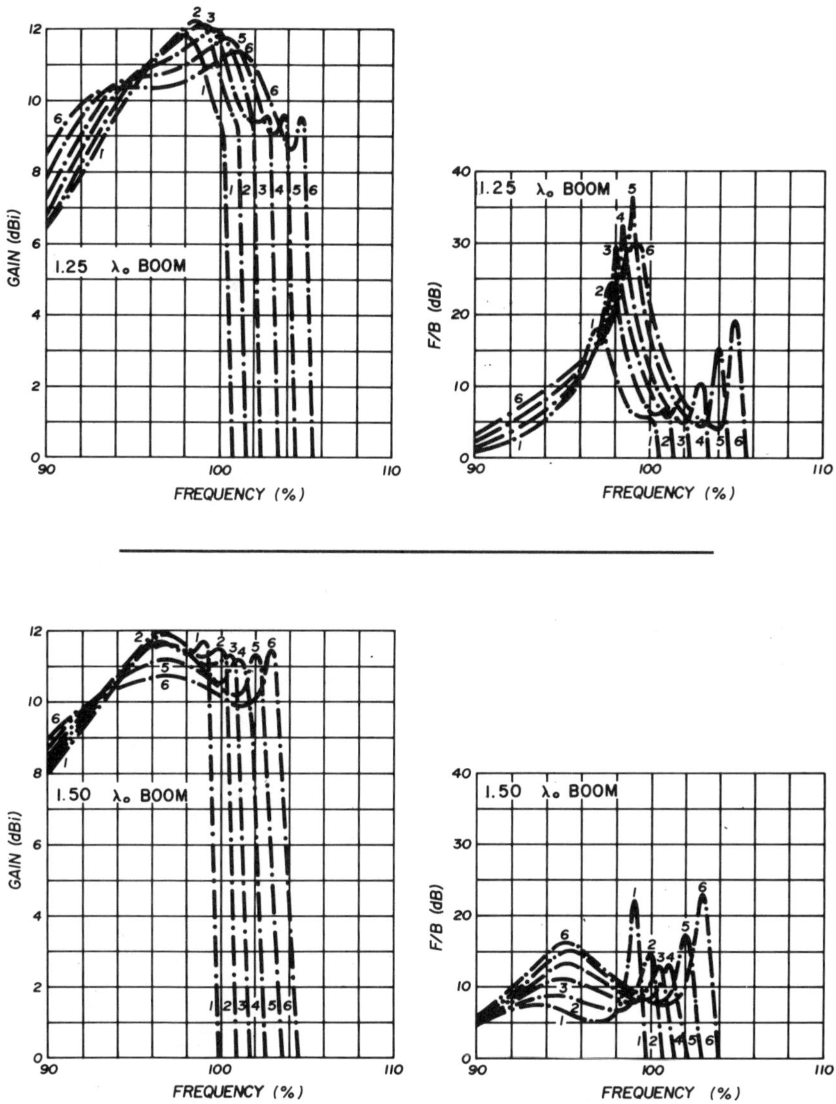

Fig. 2.11 (continued)—Gain and front-to-back ratio for six-element Yagi beams with different boom lengths and parasitic element lengths (See Table 2.3).

Performance Characteristics

If we look carefully at one of these plots, for example, Fig. 2.9 for three elements, boom = 0.25λ, it becomes clear that it is quite difficult to simply characterize "the" gain and "the" *F/B* ratio. The maximum calculated gain at a single frequency is 8.9 dBi (curve 1) but a realistic gain at the center of a practical four percent band (curve 3) is more like 8.0 dBi. Even more difficult is the characterization of the *F/B* ratio. The maximum calculated *F/B* (curve 5) is a whopping 38 dB, but this occurs only at a very specific frequency ($F = 1$) and for the situation where maximum gain is compromised (reduced to 7.3 dBi). How then can we characterize the results by a single gain figure and a meaningful *F/B* ratio?

Since gain is perhaps the most important parameter of antenna performance, and since a practical antenna must work effectively over a reasonable band, I have elected to specify the gain at the center of a four percent band. For each case, the band-center is adjusted to give maximum gain performance over the entire four percent band, and, finally the specific curve is selected which yields best overall gain performance. I define "the" gain of this case as the gain at band-center and "the" *F/B* ratio as the value at the same band-center for the same selected case. Note that the actual *F/B* may be significantly higher at some other frequency inside or outside the chosen band; we shall discuss this point shortly.

Table 2.4

Band-center gain and front to back ratio vs. boom length for various multi-element Yagi beams. Data for three, four and six elements are plotted in Figs. 2.9, 2.10 and 2.11.

	Three elements		*Four elements*		*Five elements*		*Six elements*		*Seven elements*	
boom length (λ)	*gain (dBi)*	*F/B (dB)*	*gain (dBi)*	*F/B (dB)*	*gain (dBi)*	*F/B (dB)*	*gain (dBi)*	*F/B (dB)*	*gain (dBi)*	*F/B (dB)*
0.10	6.98	12.2	7.22	12.1	7.48	8.4	7.49	7.1	7.49	6.7
0.20	7.96	14.1	7.90	5.5	7.88	7.9	7.76	5.6	7.59	4.3
0.25	8.23	19.9	8.47	8.1	8.26	10.1	8.26	7.8	7.92	5.1
0.30	9.07	12.3	8.80	7.7	8.64	9.5	8.60	8.0	8.35	6.6
0.35	9.25	11.4								
0.40	9.73	8.2	9.42	8.8	9.37	10.8	9.15	9.3	9.29	9.6
0.50	9.42	5.6	9.67	8.9	9.58	10.5	9.64	11.0	9.63	11.9
0.60	8.88	5.3	9.80	11.8	9.81	15.4	9.99	15.1	9.96	13.5
0.70	7.92	2.3	10.08	15.5	10.41	21.6	10.55	21.2	10.62	16.1
0.75			10.51	29.4	10.69	16.7	10.82	17.7	10.82	13.5
0.80			10.36	22.3	10.75	19.5	11.02	14.2	11.07	15.5
0.90			10.53	14.4	10.95	15.0	11.27	12.6	11.27	11.0
1.00			10.33	8.4	10.87	11.4	11.16	10.4	11.26	10.5
1.10			9.70	4.7	10.31	11.7	11.09	13.4	11.40	13.1
1.20					10.40	15.1	11.51	24.9	11.73	22.2
1.25					10.74	24.0	11.79	23.8	11.97	32.5
1.30					10.49	20.3	11.67	31.5	12.17	20.5
1.40					10.51	18.9	11.87	13.8	12.22	15.6
1.50					10.33	12.3	11.61	11.1	12.10	10.8

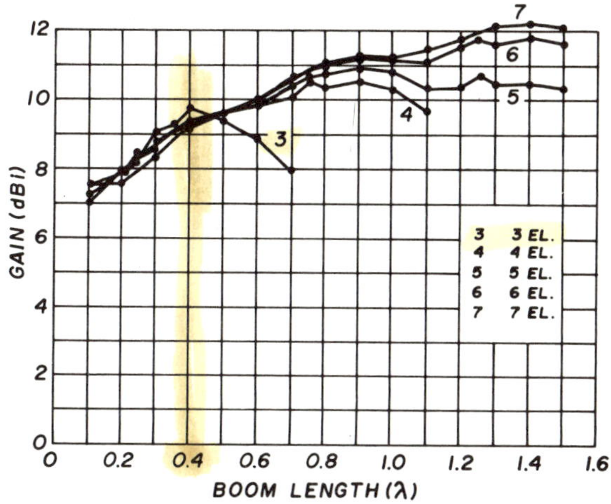

Fig. 2.12—Yagi gain for three-, four-, five-, six-, and seven-element beams as a function of boom length.

With this definition of band-center gain and band-center *F/B*, Table 2.4 has been constructed to show performance not only for three-, four-, and six-element beams, but also for five- and seven-element beams. Fig. 2.12 shows a plot of the gain information; this graph is remarkable in four respects. First of all, it demonstrates a practical upper limit to the gain achievable from a given boom length. Second, it demonstrates that this gain is almost independent of the number of elements distributed along its length as long as there are enough. Third, the achievable practical gain shows a slight preference for more rather than fewer elements on a boom. Finally, the "boom gain"—achievable gain from a given boom length—is not really a smooth function of boom length. Instead, it appears to exhibit "bumps" or oscillations with a fraction of a decibel amplitude and spacing of about a half-wavelength.

This concept of boom gain, independent of the number of elements is not new; in fact, it was suggested by Ehrenspeck and Poehler in a series of experiments using the automatic plotter built at the Air Force Cambridge Research Center.[4] No claims by Ehrenspeck and Poehler were made as to the absolute accuracy of their results, but they were able to demonstrate essential independence of gain on number of elements over rather wide limits for two long Yagi models (1.2λ and 6.0λ). If one accepts the idea of universal boom gain, it is instructive to compare the (upper envelope) curve of Fig. 2.12 with Ehrenspeck and Poehler's experimental points, as well as the experimental results of Lindsay.[5] Lindsay made a number of models of varying boom length (but unstated element dimension schedules) and measured directivity at a design frequency of 440 MHz. All of these results are shown in Fig. 2.13 where the solid curve is the theoretical maximum gain

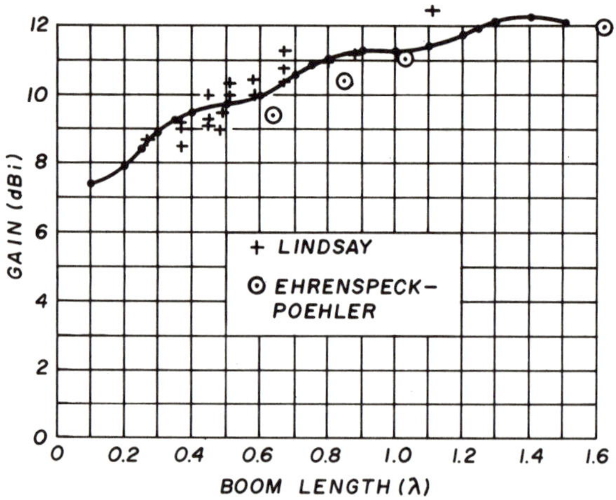

Fig. 2.13—Theoretical Yagi gain envelope vs. boom length compared with experimental data of Ehrenspeck-Poehler and Lindsay.

(from Fig. 2.12) and the keyed points are from Ehrenspeck-Poehler and from Lindsay. The Lindsay experiments provide remarkable confirmation of the universal boom gain curve. The Ehrenspeck-Poehler measurements all appear to lie slightly below the theoretical curve (by a fraction of a decibel). It is not clear that the slight discrepancy in absolute value is a real disagreement; it may be within the expected accuracy of the gain calibration technique used on the automatic plotter. It may also be due to lack of optimization; Ehrenspeck and Poehler used a fixed reflector reactance and it is hard to guess how much more gain they would have found with an optimized configuration.

Element Illumination

Fig. 2.14 taken from Table 2.4 shows a plot of the band-center *F/B* ratio as a function of overall length. It is notable that there are three empirical values of overall length which seem to produce high values of *F/B* independent of the number of elements. These apparently favorable overall lengths are 0.25λ, 0.75λ, and 1.25λ—all odd multiples of a quarter-wave. For the 0.25λ position only the three-element beam shows a high value, but this is primarily caused by the definition of band-center *F/B* ratio.

This remarkable phenomenon suggests that there might be a basic physical explanation covering all cases; indeed, such a physical basis is not hard to find. Analogous to the optical illumination of an aperture by light, one can think of the Yagi boom length as illuminated by electrical excitation. Unlike the case of uniform illumination of an optical aperture, the Yagi illumination is not uniform but can be viewed as a series of discrete excitation points (the

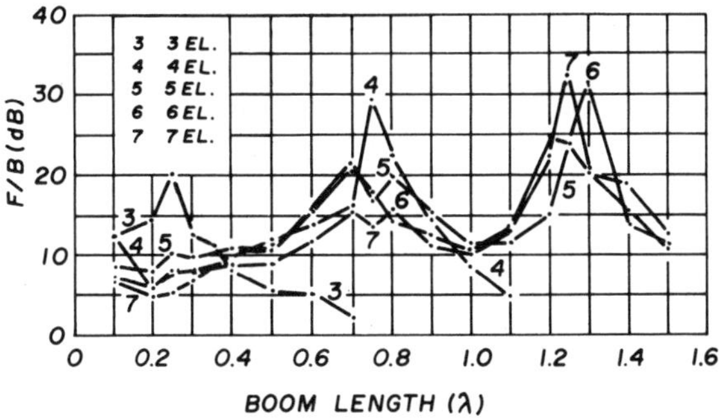

Fig. 2.14—Band-center front-to-back ratio for three-, four-, five-, six-, and seven-element Yagis vs. boom length.

elements) whose average envelope is quasi-uniform. In the optical case the wave front is ordinarily plane (phase shift across the aperture is zero), whereas in the Yagi case the phase shift is purposely designed to cause the main diffracted "beam" to lie along the boom rather than broadside to the aperture as in the optical case.

The aperture produces a diffraction pattern (beam pattern) consisting of a "main beam" and several lobes. The number of lobes is determined basically by the size of the aperture in wavelengths, while the amplitude of the lobes is determined by the way the aperture is illuminated (phase and amplitude).

An informative treatment of an end-fire array (aperture illuminated by a series of radiators having equal amplitudes) is given in Kraus (ref. 3, pages 76-89). There are two interesting cases: the ordinary end-fire array in which the angular phase change between radiators is just equal to their spatial separation angle, and the increased directivity end-fire array first derived by Hansen and Woodyard where the radiator phase delay is larger than spatial separation by an angle π/n.[6] The latter case is a good one with which to compare the Yagi antenna, because although the amplitude of the current(s) in the Yagi elements are not uniform, the phases are adjusted (by element reactance) to give highest directivity or gain. Kraus gives expected null angle directions for these two cases (null between lobes) as:

Ordinary end-fire:

$$\theta_0 = 2 \sin^{-1} [\pm (m\lambda/2ns)^{1/2}] \tag{2.1}$$

Increased directivity end-fire:

$$\theta_0 = 2 \sin^{-1} [\pm((2m - 1)\lambda/4ns)^{1/2}] \tag{2.2}$$

In our model the boom length l_B is $(n - 1)s$ so that we can rewrite these equations as:

Ordinary end-fire:

$$\theta_0 = 2 \sin^{-1} [\pm(m(n - 1)\lambda/2l_B n)^{1/2}] \tag{2.3}$$

Increased-directivity end-fire:

$$\theta_0 = 2 \sin^{-1} [\pm((2m-1)(n-1)\lambda/4l_B n)^{1/2}] \qquad (2.4)$$

Note that where the number of elements, n, is large one would expect a high F/B ratio (that is, a null at 180 degrees) for particular values of boom length essentially independent of the number of elements.

Let's now examine the patterns of the cases where F/B is relatively high for the six-element beam at three boom lengths: 0.25λ, 0.75λ, and 1.25λ at the precise frequency where the F/B ratio is maximum (presumably where the back radiation "null" occurs). It is instructive to plot the pattern not only at this "best" frequency but also at frequencies just below and just above the best frequency. For all of these cases the reflector length is fixed at 0.50702λ (F = 0.95), and all director lengths are fixed at 0.45873λ (F = 1.05).

Fig. 2.15 shows the H- and E-plane patterns for all of these cases. The H-plane pattern shows "nulls" between lobes which move with frequency (equivalent to changing boom length in actual wavelengths). We can compare the angle at which these nulls occur to those of the end-fire arrays from (2.3) and (2.4); Table 2.5 lists these comparisons.

Note that these comparisons show qualitative agreement; also note that the computed Yagi results are in better agreement with the increased-directivity end-fire model. The more rapid shift of null angle(s) with frequency for the Yagis compared to either end-fire model is not to be taken too seriously because, as we shift frequency, not only does the effective boom length in terms of actual wavelength change, but the element reactance(s) change significantly, i.e., the Yagi really becomes a different Yagi.

The details of the Yagi pattern depend on the particular way in which the boom is illuminated, i.e., on the details of element positions, and element current magnitudes and phases. The depth of the nulls depends on the degree of vectorial cancellation of back radiation; since the vectors themselves vary significantly with all Yagi parameters it is no wonder that complete cancellation is accidental and ordinarily impossible. The size of the lobes is determined primarily by the shape and phase delay of the Yagi illumination function. For the uniformly illuminated case the reader is referred to the uniform end-fire arrays (see Kraus, pages 79-88).

It will be noted that for non-uniformly illuminated broadside structures (Kraus, pages 93-121) the side-lobe level is highest for "edge" illumination, next highest for uniform illumination, and zero for illumination based on element amplitudes following the coefficients of a binomial series. One can expect the same kind of result for an end-fire array where edge illumination (two-element beam) produces high sidelobes, uniform illumination smaller sidelobes, and an illumination function falling at the extreme edges (reflector and end director currents smaller than in central elements) produces still smaller sidelobes.

Unfortunately, for a given overall boom length the directivity suffers somewhat as illumination is adjusted for smaller sidelobes. Moreover, in the

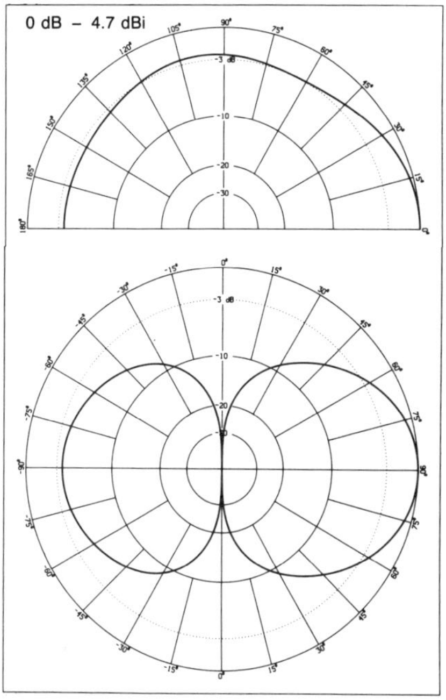

F = 0.95

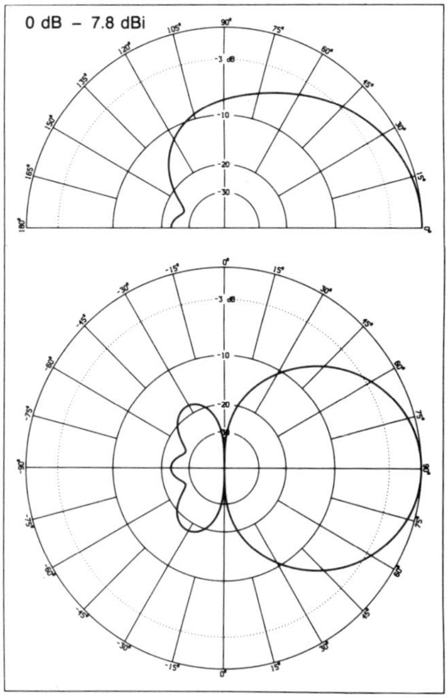

F = 1.00

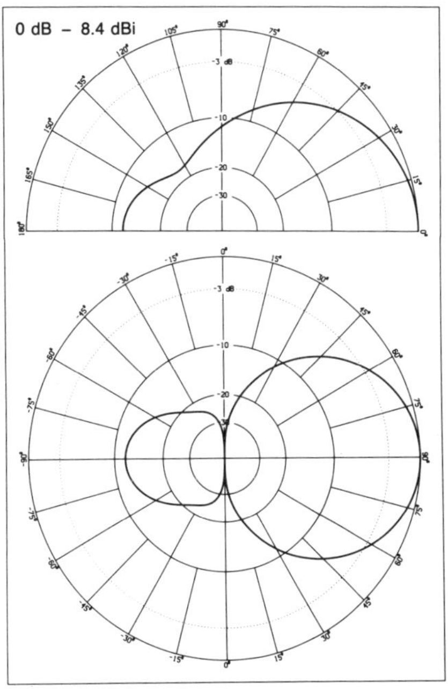

F = 1.05

Fig. 2.15—E- and H-plane patterns for a six-element Yagi with 0.25 wavelength boom.

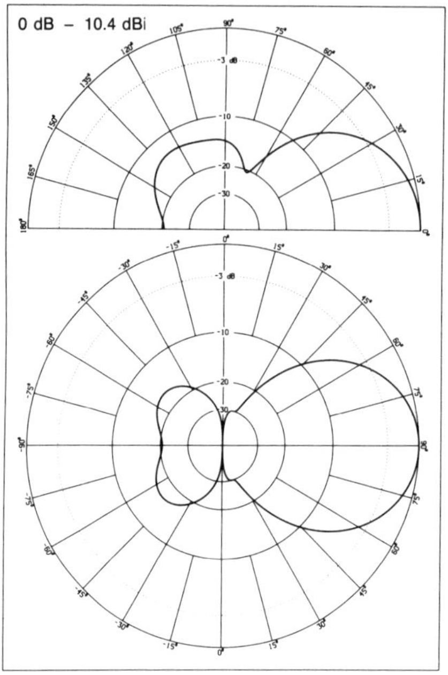

F = 0.95

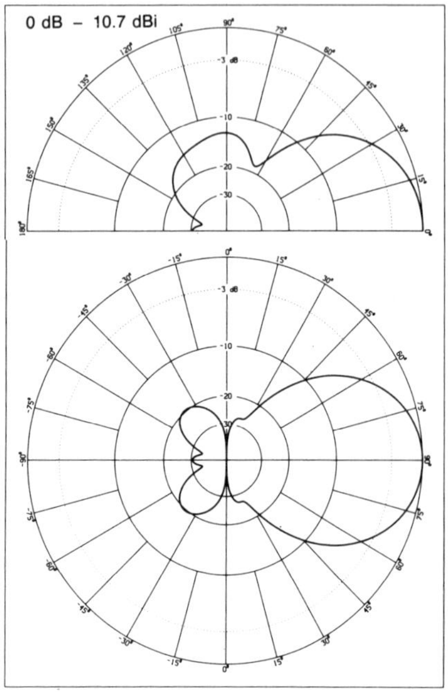

F = 1.00

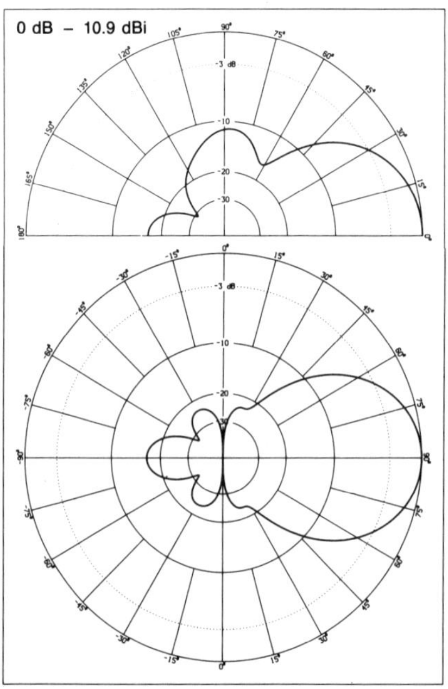

F = 1.05

Fig. 2.15 (continued)—E- and H-plane patterns for a six-element Yagi with 0.75 wavelength boom.

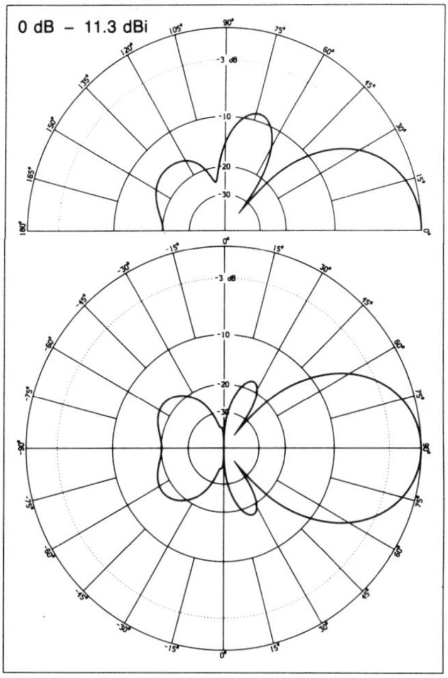

F = 0.95

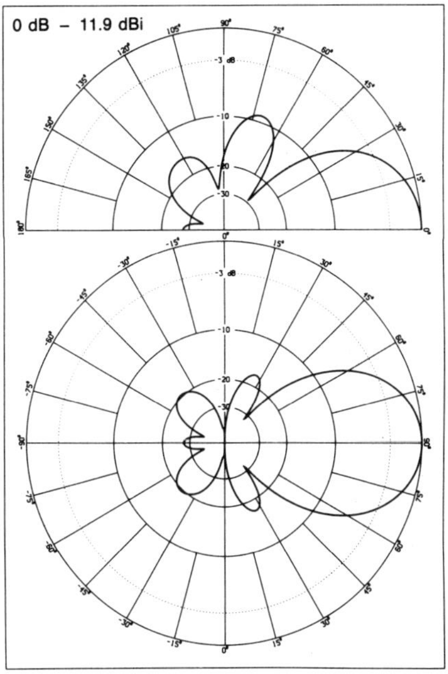

F = 1.00

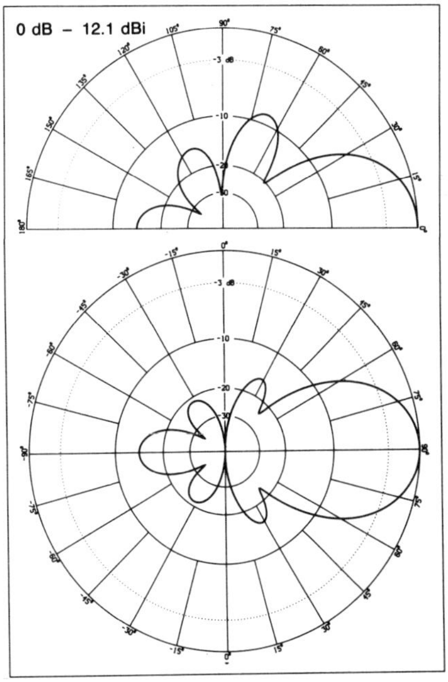

F = 1.05

Fig. 2.15—(continued)—E- and H-plane patterns for a six-element Yagi with 1.25 wavelength boom.

Table 2.5

Yagi null angles compared to ordinary end-fire arrays (OEF) and increased directivity end-fire (IDEF).

boom length (λ)	*freq*	*m = 1 null (degrees)* Yagi	OEF	IDEF	*m = 2 null (degrees)* Yagi	OEF	IDEF	*m = 3 null (degrees)* Yagi	OEF	IDEF
0.25	0.972	—								
0.25	0.982	159								
0.25	0.992	120								
0.75	0.970	69	98	65	—	—	136			
0.75	0.980	66	98	64	171	—	135			
0.75	0.990	63	97	64	147	—	133			
1.25	0.976	51	72	49	102	112	91	—	—	135
1.25	0.986	51	71	49	99	111	91	171	—	134
1.25	0.996	48	71	48	93	110	90	150	—	132

case of a simplistic Yagi the illumination function is hard to adjust; the current amplitudes and phases are all determined by element reactances and positions and it is necessary to simply accept the result. Table 2.6 shows the current amplitudes and phases in each element for the case of the six-element beam, boom = 0.75λ where the driver current is set at 1.0 ampere at 0 degrees phase. Note that while the current amplitudes are not "uniform illumination," they seem to average out surprisingly alike. Incidentally, the current in the last director is always characteristically higher than that in the preceding director; this is due to the "end effect" (no mutual coupling to an element ahead of it).

Table 2.6

Element currents for a six-element Yagi with a boom length of 0.75 λ at F = 0.980.

element	*current (amps)*	*phase (degrees)*
Reflector	0.476	154.5
Driven element	1.000	0.0
Director 1	0.379	−121.3
Director 2	0.467	−169.7
Director 3	0.258	115.3
Director 4	0.458	32.6

Thus we now have a consistent picture of the high *F/B* ratio Yagi design; the essence of correct design is to place the null between lobes exactly in the back direction. This will occur for simplistic Yagi antennas when the overall length is approximately an odd multiple of a quarter-wavelength. The specific best design will involve optimizing the boom length and the boom illumination function (the particular element excitation currents and phases) to yield the best *F/B* ratio. Such optimization can be carried out around the best boom lengths; this is discussed in Chapter 3. For the present it is sufficient to note that really excellent *F/B* ratios are possible with these simplistic designs as long as one is willing to accept boom lengths which are approximately odd multiples of a quarter-wavelength.

As a final note on these simplistic best designs, not only is the back radiation

Table 2.7
Sidelobe ratios for six-element Yagis with various boom lengths.

boom length (λ)	*freq*	*F/B* (dB)	*E-plane peak sidelobe* (dB)	*H-plane peak sidelobe* (dB)
0.25	0.982	25.7	18.8	7.5
0.75	0.980	32.5	19.7	12.6
1.25	0.986	35.0	18.8	14.9

low but the minimum (worst case) front-to-sidelobe ratio in the entire reverse direction (90 to 180 degrees) is also surprisingly high. Table 2.7 shows the result. These Yagi designs seem to be generally excellent.

With the exception of the two-element beam case I have not yet commented on the driving-point impedance of any of the Yagis shown. Remember that one can, by adjusting the length of the driver, always null out the driving-point reactance at a designated frequency. The remaining resistance, however, just like that for the two-element beam, varies enormously from case to case. It is very low for very short beams where there is very strong coupling between elements; moreover, in such cases it varies wildly with frequency as does the change in reactance with frequency. Thus to ensure a reasonably reliable electrical feed system it is wise to keep element separation well above 0.05λ; all such cases investigated have reasonably well behaved driving-point impedances.

Summary

Let me summarize the results for simplistic Yagi antennas:

1) Two- to seven-element beams with boom lengths to 1.5λ have been systematically explored.

2) Simplistic Yagis display a gain function where bandwidth is primarily a function of the resonant frequency separation between reflector and directors. The bandwidth can easily be made several percent of the central frequency.

3) The shape of the gain function is generally not flat in the region of interest. It is also more complex for beams which use a large number of elements.

4) Simplistic Yagis display a *F/B* ratio function with a shape that varies enormously from case to case. The shape may contain more than one peak and changes rapidly with boom length and/or frequency. It is so complicated that it is not possible to characterize its bandwidth.

5) High values of the *F/B* ratio (more than 30 dB) are quite rare; when they occur *F/B* is high only over a very narrow band of frequencies.

6) The spacing between elements should be generally greater than 0.05λ to realize a well-behaved feed.

7) The maximum practical gain of the simplistic Yagi is almost entirely determined by boom length. Maximum gain increases, but not steadily, with boom length.

8) Best design for a high *F/B* ratio requires the approximate boom length to be an odd multiple of a quarter-wavelength at the design frequency.

NOTES:

[1]Hall, G. L. ed., *The ARRL Antenna Book*, 13th Edition, Newington, Connecticut: American Radio Relay League, 1974.

[2]Brown, G. H., "Directional Antennas," *Proceedings of the IRE*, January 1937.

[3]Kraus, John D., *Antennas*, New York: McGraw-Hill, 1950.

[4]Ehrenspeck, H. and Poehler, H., "A New Method of Obtaining Maximum Gain from Yagi Antennas," *IRE Transactions on Antennas and Propagation*, October 1959.

[5]Lindsay, J., "Quads and Yagis," *QST*, May 1968.

[6]Hansen, W. W. and Woodyard, J. R., "A New Principle in Directional Antenna Design," *Proceedings of the IRE*, March 1938.

CHAPTER 3

YAGI ANTENNA PERFORMANCE OPTIMIZATION

In Chapter 2, I explored the properties of simplistic Yagi antennas; that is, antennas of a given boom length but having uniformly spaced elements, one of which is a reflector, one of which is a driver, and with director parasites all of uniform length. In reality, however, one is not restricted to the simplistic design. It is therefore interesting to examine a number of departures from the simplistic design to see if there are ways to further improve performance. In this chapter I explore some ideas in a systematic way. It will soon be apparent that some of the departures from the simplistic design produce only subtle changes in performance, while others produce significant changes in performance. It is fortunate that the accuracy inherent in computation can show very subtle changes; these changes, although small, can usually be trusted. However, the absolute accuracy of the model on which computation is based may not be better than a few percent.

Departures from the simplistic design may be made in a number of ways but primarily by allowing the lengths of the directors (hence their resonant frequencies) to vary and by allowing the placement of the elements on the boom to change. Additionally, for a given boom length and a given total number of elements, the number of reflectors (and hence directors) can be changed. This is a much more drastic change and produces more pronounced performance variations. These changes are analyzed. Only free space performance is investigated at this time.

I shall start with a "good" simplistic design consisting of a six-element Yagi on a boom 0.75λ long, but will first change the lengths of all parasites to bring the center frequency of the desired 4-percent band of maximum gain to $F = 1.0$. Fig. 3.1 shows the main properties of this test antenna and the required element lengths. Note that the position of maximum F/B ratio is somewhat lower ($F = 0.988$) than the gain band center ($F = 1.0$); Fig. 3.2 shows the pattern of this antenna at band center. Two visible nulls can be seen. The first one occurs at 63 degrees and the second at 144 degrees. The second null can be identified with the peak in the F/B ratio occurring at $F = 0.988$; at this lower frequency the null moves out to 180 degrees. This antenna can be operated at best gain over the band and compromise F/B (17 dB at $F = 1.0$), or operated at the frequency of best F/B ($F = 0.988$, F/B = 38 dB) and somewhat compromise the gain available. In either case this Yagi design seems to be a good one and I shall use it as a test case around which certain departures from simplistic design can take place.

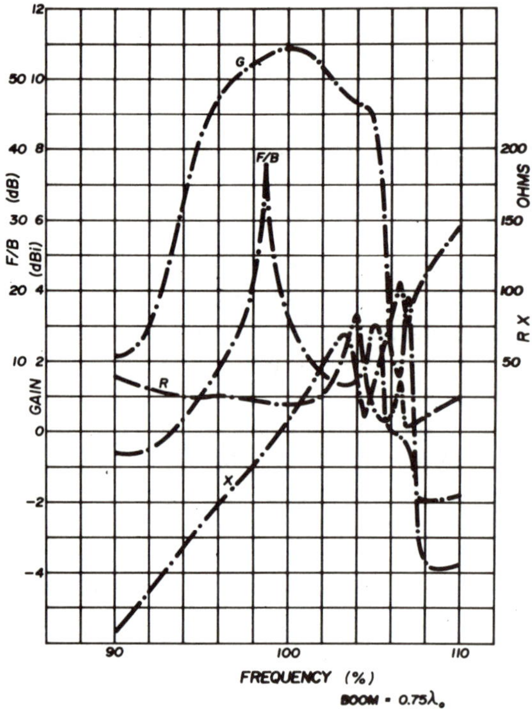

element	length (λ_o)	free space resonance
Reflector	0.50195	0.96
Driven Element	0.48167	1.00
Director 1	0.45414	1.06
Director 2	0.45414	1.06
Director 3	0.45414	1.06
Director 4	0.45414	1.06

Fig. 3.1 — Performance characteristics of a six-element Yagi beam with a boom length of 0.75λ.

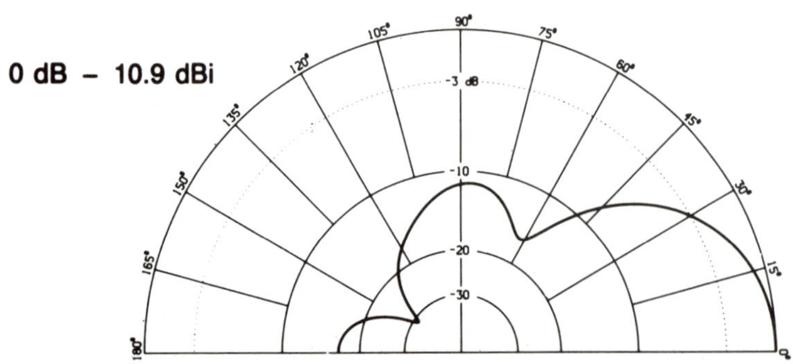

Fig. 3.2 — Radiation pattern of a six-element Yagi beam on a 0.75λ boom. The pattern nulls occur at 63 and 144 degrees; the second null is correlated with the peak of *F/B* ratio at *F* = 0.988.

Table 3.1 — Computed performance characteristics of the six-element Yagi over the frequency range from *F* = 0.970 to *F* = 1.030.

frequency (F)	*gain (dBi)*	*F/B ratio (dB)*	*feedpoint resistance (ohms)*	*feedpoint reactance (ohms)*
0.970	10.06	12.9	24.7	−38.2
0.972	10.15	14.0	24.5	−35.5
0.974	10.23	15.3	24.2	−32.8
0.976	10.30	16.7	23.9	−30.0
0.978	10.37	18.3	23.6	−27.2
0.980	10.43	20.2	23.2	−24.3
0.982	10.50	22.7	22.8	−21.4
0.984	10.55	26.0	22.5	−18.5
0.986	10.61	31.2	22.1	−15.5
0.988	10.66	38.0	21.7	−12.4
0.990	10.71	31.8	21.4	−9.2
0.992	10.75	26.5	21.1	−6.0
0.994	10.79	23.0	20.8	−2.7
0.996	10.82	20.6	20.5	0.6
0.998	10.84	18.6	20.3	4.1
1.000	10.86	17.0	20.2	7.5
1.002	10.86	15.7	20.2	11.1
1.004	10.86	14.5	20.2	14.8
1.006	10.85	13.5	20.4	18.5
1.008	10.82	12.6	20.7	22.3
1.010	10.78	11.7	21.1	26.2
1.012	10.73	11.0	21.7	30.2
1.014	10.67	10.3	22.5	34.3
1.016	10.59	9.7	23.5	38.4
1.018	10.50	9.1	24.8	42.6
1.020	10.39	8.6	26.5	46.9
1.022	10.28	8.1	28.6	51.2
1.024	10.16	7.7	31.3	55.5
1.026	10.04	7.4	34.7	59.8
1.028	9.91	7.1	39.1	63.8
1.030	9.79	6.9	44.7	67.3

Since we are interested in very subtle changes I show in Table 3.1 the detailed performance in the region of chief interest, accurately calculated for this antenna over the frequency range from $F = 0.970$ to $F = 1.030$. The driving point reactance at a given frequency is somewhat arbitrary; it can be easily shifted or offset by changing the length of the driven element. The length of the driven element, however, remains fixed (free space resonant frequency = 1.0) throughout this series of explorations so that reactance changes can be properly sensed.

Parasite Length Variations

If we keep the average value of the lengths of the directors constant we can explore "linear" length tapers and "parabolic" tapers. I will define a linear taper as a uniform linear free space resonant frequency progression from D1 to D4 keeping the average resonant frequency constant. In other words, the director resonant frequencies are linearly related to director position measured from the center of the director assembly (to keep the average value constant). Similarly, a "parabolic" taper is one in which the directors' free space resonant frequencies are proportional to the square of the distance from the center of the director assembly, with the further condition that the average value of director resonance is held constant. If we define the change in resonant frequency of D1 as Δ, then with a total of four directors all other director resonances will change:

element	*linear taper*	*parabolic taper*
Director 1	$+\Delta$	$+\Delta$
Director 2	$+\Delta/3$	$-\Delta$
Director 3	$-\Delta/3$	$-\Delta$
Director 4	$-\Delta$	$+\Delta$

The degree or magnitude of taper is fixed by the size of Δ; moreover, Δ can be chosen either as an increase or decrease in resonant frequency.

I have selected and investigated six taper schedules which are delineated in Table 3.2; as indicated, antenna performance in the critical region of interest is shown in Tables 3.3 to 3.8. These results are to be compared with the simplistic design shown in Table 3.1.

Table 3.3 shows the results for a "−2 percent" linear taper and it is obvious that all performance parameters are virtually unchanged. Table 3.4 shows the results for a "−4 percent" linear taper, and even in this rather extreme case performance is almost totally unchanged in the central frequency region. Remember that for this case the free space resonant frequency of D1 has dropped from 1.06 to 1.02 and it is easy to see that performance deteriorates at higher frequencies (D1 no longer behaves like a director). Nevertheless, it is remarkable that the linear taper — even of this magnitude — has virtually no effect on the performance of the Yagi antenna.

Table 3.5 shows results for a −2 percent parabolic taper; Tables 3.6, 3.7 and 3.8 show results for parabolic tapers of −1, +1, and +2 percent, respectively. In comparing these tables with the standard simplistic Yagi antenna we again see that truly minimal changes are made to the chief performance indices. The variations in maximum *F/B* ratios are not of great significance; I shall come back to this point later.

Table 3.2 — Schedule of director lengths for the six six-element Yagi-Uda performance characteristics listed in Table 3.3 through Table 3.8.

Table	*taper type*	*Dir 1 (λ)*	*Dir 2 (λ)*	*Dir 3 (λ)*	*Dir 4 (λ)*	Δ
3.3	Linear	0.46341	0.45719	0.45263	0.44524	−2%
3.4	Linear	0.47306	0.46028	0.44817	0.43667	−4%
3.5	Parabolic	0.46341	0.44524	0.44524	0.46341	−2%
3.6	Parabolic	0.45873	0.44964	0.44964	0.45873	−1%
3.7	Parabolic	0.44964	0.45873	0.45873	0.44964	+1%
3.8	Parabolic	0.44524	0.46341	0.46341	0.44524	+2%

Table 3.3 — Performance characteristics of a six-element Yagi beam with a boom length of 0.75λ, director lengths tapered linearly at −2 percent (director lengths shown in Table 3.2).

frequency (F)	*gain (dBi)*	*F/B ratio (dB)*	*feedpoint resistance (ohms)*	*feedpoint reactance (ohms)*
0.970	10.07	13.0	23.9	−39.1
0.972	10.16	14.1	23.6	−36.4
0.974	10.24	15.3	23.4	−33.7
0.976	10.31	16.6	23.1	−31.0
0.978	10.38	18.2	22.8	−28.3
0.980	10.44	20.0	22.4	−25.5
0.982	10.49	22.3	22.1	−22.6
0.984	10.54	25.3	21.7	−19.8
0.986	10.59	29.9	21.4	−16.8
0.988	10.64	40.0	21.0	−13.9
0.990	10.67	38.2	20.7	−10.9
0.992	10.71	29.4	20.4	−7.8
0.994	10.74	25.1	20.1	−4.7
0.996	10.76	22.2	19.9	−1.5
0.998	10.77	20.1	19.7	1.7
1.000	10.78	18.4	19.5	4.9
1.002	10.78	16.9	19.5	8.3
1.004	10.77	15.7	19.5	11.6
1.006	10.75	14.6	19.6	15.0
1.008	10.72	13.7	19.8	18.4
1.010	10.68	12.9	20.1	21.8
1.012	10.63	12.1	20.6	25.3
1.014	10.57	11.5	21.2	28.7
1.016	10.50	10.9	22.0	32.1
1.018	10.42	10.3	23.0	35.5
1.020	10.33	9.9	24.2	38.9
1.022	10.24	9.5	25.7	42.0
1.024	10.15	9.2	27.6	45.0
1.026	10.05	8.9	29.9	47.6
1.028	9.95	8.8	32.6	49.6
1.030	9.86	8.7	35.6	50.7

Table 3.4 — Performance characteristics of a six-element Yagi beam with a boom length of 0.75λ, director lengths tapered linearly at −4 percent (director lengths shown in Table 3.2).

frequency (F)	*gain (dBi)*	*F/B ratio (dB)*	*feedpoint resistance (ohms)*	*feedpoint reactance (ohms)*
0.970	10.06	13.7	20.9	−40.3
0.972	10.15	14.9	20.6	−37.5
0.974	10.23	16.2	20.2	−34.7
0.976	10.31	17.7	19.9	−31.9
0.978	10.37	19.5	19.5	−29.1
0.980	10.43	21.6	19.1	−26.2
0.982	10.49	24.2	18.6	−23.3
0.984	10.54	27.8	18.2	−20.3
0.986	10.58	31.9	17.8	−17.3
0.988	10.61	31.9	17.5	−14.3
0.990	10.64	27.7	17.1	−11.3
0.992	10.66	24.3	16.8	−8.2
0.994	10.68	21.7	16.5	−5.1
0.996	10.68	19.6	16.2	−1.9
0.998	10.67	17.9	16.0	1.3
1.000	10.65	16.5	15.8	4.5
1.002	10.62	15.3	15.7	7.7
1.004	10.58	14.2	15.6	10.9
1.006	10.52	13.2	15.6	14.1
1.008	10.45	12.4	15.7	17.3
1.010	10.37	11.6	15.8	20.5
1.012	10.26	10.9	16.1	23.6
1.014	10.15	10.2	16.4	26.7
1.016	10.01	9.7	16.7	29.7
1.018	9.86	9.1	17.1	32.5
1.020	9.69	8.6	17.6	35.2
1.022	9.49	8.2	18.0	37.6
1.024	9.25	7.7	18.2	39.8
1.026	8.96	7.3	18.2	41.8
1.028	8.56	6.7	17.7	43.7
1.030	8.01	6.0	16.7	45.8

Table 3.5 — Performance characteristics of a six-element Yagi beam with a boom length of 0.75λ, director lengths tapered parabolically at −2 percent (director lengths shown in Table 3.2).

frequency (F)	*gain (dBi)*	*F/B ratio (dB)*	*feedpoint resistance (ohms)*	*feedpoint reactance (ohms)*
0.970	10.01	16.4	19.7	−42.3
0.972	10.11	18.1	19.2	−39.4
0.974	10.21	20.2	18.6	−36.4
0.976	10.30	22.8	18.1	−33.3
0.978	10.39	26.2	17.5	−30.3
0.980	10.47	29.7	17.0	−27.1
0.982	10.55	29.3	16.4	−23.9
0.984	10.63	25.7	15.9	−20.7
0.986	10.71	22.4	15.4	−17.4
0.988	10.78	19.9	15.0	−14.0
0.990	10.84	17.9	14.6	−10.6
0.992	10.90	16.3	14.2	−7.1
0.994	10.95	14.8	13.9	−3.6
0.996	10.99	13.6	13.7	0.0
0.998	11.01	12.5	13.5	3.7
1.000	11.03	11.5	13.4	7.4
1.002	11.03	10.7	13.3	11.1
1.004	11.02	9.9	13.4	15.0
1.006	10.99	9.1	13.6	18.9
1.008	10.95	8.4	13.9	22.8
1.010	10.89	7.8	14.3	26.9
1.012	10.82	7.3	14.9	31.0
1.014	10.74	6.7	15.7	35.2
1.016	10.64	6.3	16.8	39.5
1.018	10.54	5.8	18.1	44.0
1.020	10.44	5.5	20.0	48.5
1.022	10.34	5.2	22.3	53.2
1.024	10.24	4.9	25.5	57.9
1.026	10.16	4.7	30.0	62.7
1.028	10.10	4.6	36.2	67.0
1.030	10.07	4.6	45.3	69.9

Table 3.6 — Performance characteristics of a six-element Yagi beam with a boom length of 0.75λ, director lengths tapered parabolically at −1 percent (director lengths shown in Table 3.2).

frequency (F)	*gain (dBi)*	*F/B ratio (dB)*	*feedpoint resistance (ohms)*	*feedpoint reactance (ohms)*
0.970	10.04	14.3	22.5	−40.5
0.972	10.13	15.7	22.1	−37.7
0.974	10.22	17.2	21.7	−34.9
0.976	10.30	19.0	21.3	−32.0
0.978	10.38	21.2	20.8	−29.1
0.980	10.45	23.9	20.3	−26.1
0.982	10.52	27.6	19.8	−23.0
0.984	10.59	31.6	19.4	−20.0
0.986	10.66	30.4	18.9	−16.8
0.988	10.72	26.3	18.5	−13.6
0.990	10.77	22.9	18.1	−10.3
0.992	10.82	20.4	17.7	−7.0
0.994	10.87	18.4	17.4	−3.7
0.996	10.91	16.8	17.1	−0.1
0.998	10.94	15.4	16.9	3.4
1.000	10.96	14.2	16.8	7.0
1.002	10.97	13.1	16.7	10.7
1.004	10.97	12.2	16.8	14.4
1.006	10.96	11.3	16.9	18.3
1.008	10.93	10.5	17.2	22.2
1.010	10.89	9.8	17.6	26.1
1.012	10.84	9.2	18.1	30.2
1.014	10.77	8.6	18.9	34.3
1.016	10.69	8.0	19.9	38.6
1.018	10.61	7.5	21.2	42.9
1.020	10.51	7.1	22.9	47.4
1.022	10.41	6.7	25.0	51.9
1.024	10.31	6.4	27.9	56.5
1.026	10.22	6.1	31.6	61.1
1.028	10.13	5.9	36.6	65.5
1.030	10.05	5.7	43.4	69.4

Table 3.7 — Performance characteristics of a six-element Yagi beam with a boom length of 0.75λ, director lengths tapered parabolically at +1 percent (director lengths shown in Table 3.2).

frequency (F)	*gain (dBi)*	*F/B ratio (dB)*	*feedpoint resistance (ohms)*	*feedpoint reactance (ohms)*
0.970	10.07	11.9	26.2	−35.8
0.972	10.15	12.9	26.1	−33.1
0.974	10.23	13.9	26.0	−30.4
0.976	10.29	15.1	25.9	−27.7
0.978	10.36	16.4	25.7	−25.0
0.980	10.41	17.9	25.5	−22.2
0.982	10.47	19.7	25.2	−19.4
0.984	10.51	21.9	25.0	−16.5
0.986	10.56	24.9	24.7	−13.6
0.988	10.60	29.3	24.4	−10.6
0.990	10.64	38.4	24.1	−7.6
0.992	10.67	38.8	23.9	−4.5
0.994	10.69	29.4	23.7	−1.3
0.996	10.71	25.0	23.5	2.0
0.998	10.73	22.1	23.3	5.3
1.000	10.73	19.9	23.2	8.8
1.002	10.73	18.1	23.2	12.3
1.004	10.71	16.7	23.3	15.9
1.006	10.69	15.4	23.5	19.5
1.008	10.65	14.3	23.8	23.3
1.010	10.60	13.3	24.3	27.1
1.012	10.53	12.5	24.9	31.0
1.014	10.45	11.7	25.8	35.0
1.016	10.35	10.9	26.8	39.1
1.018	10.24	10.3	28.2	43.2
1.020	10.11	9.7	29.9	47.3
1.022	9.97	9.2	32.0	51.5
1.024	9.81	8.7	34.5	55.5
1.026	9.64	8.3	37.7	59.5
1.028	9.46	7.9	41.6	63.1
1.030	9.28	7.6	46.4	66.3

Table 3.8 — Performance characteristics of a six-element Yagi beam with a boom length of 0.75λ, director lengths tapered parabolically at +2 percent (director lengths shown in Table 3.2).

frequency (F)	*gain (dBi)*	*F/B ratio (dB)*	*feedpoint resistance (ohms)*	*feedpoint reactance (ohms)*
0.970	10.07	11.0	27.1	−33.5
0.972	10.15	11.9	27.2	−30.8
0.974	10.22	12.9	27.2	−28.2
0.976	10.28	13.9	27.2	−25.5
0.978	10.33	15.1	27.2	−22.8
0.980	10.38	16.4	27.1	−20.1
0.982	10.43	17.8	26.9	−17.3
0.984	10.47	19.5	26.8	−14.5
0.986	10.50	21.6	26.6	−11.6
0.988	10.53	24.2	26.4	−8.7
0.990	10.56	27.8	26.1	−5.7
0.992	10.58	32.7	25.9	−2.6
0.994	10.59	33.6	25.7	0.5
0.996	10.59	28.7	25.6	3.7
0.998	10.59	24.8	25.4	7.1
1.000	10.58	22.0	25.3	10.5
1.002	10.56	19.8	25.3	14.0
1.004	10.53	18.1	25.4	17.5
1.006	10.48	16.6	25.6	21.2
1.008	10.41	15.3	25.9	25.0
1.010	10.33	14.1	26.3	28.9
1.012	10.23	13.1	27.0	32.9
1.014	10.11	12.2	27.8	36.9
1.016	9.92	11.3	28.8	41.0
1.018	9.79	10.6	30.1	45.2
1.020	9.60	9.8	31.7	49.5
1.022	9.38	9.2	33.6	53.7
1.024	9.14	8.5	36.0	58.0
1.026	8.87	7.9	38.9	62.2
1.028	8.57	7.4	42.3	66.2
1.030	8.24	6.8	46.4	69.9

Parasite Placement Variations

Analogous to the length (and corresponding resonant frequency) taper variations I have also investigated element placement variations along the boom. Again it is possible to vary the space intervals between elements linearly and pseudo-parabolically. Table 3.9 shows element positions for six schedules I have investigated and the individual results are shown in Tables 3.10 through 3.15.

Table 3.10 shows the results where elements are crowded towards D4. Note that truly large changes in placement have been made. Similarly, Table 3.13 shows the results where elements are severely crowded towards the reflector. Tables 3.11 and 3.12 are intermediate schedules; Tables 3.14 and 3.15 show results where end spaces are (relatively) increased. These tables all show that these placement variations have only a very minor effect on directivity, and while the maximum *F/B* ratio is somewhat affected (generally adversely), we shall soon see that this effect may not be very significant.

Up to this point we have looked at taper schedules which are linear and parabolic and which also involve director length, or resonant frequency, and element placement along the boom. It is truly remarkable that all of these schedules produce minimal changes in antenna performance; it is therefore plausible that combinations of these schedules will also produce minimal performance variations. This leads to the conclusion that the original simplistic design (dimensions listed in Fig. 3.1) is just about as good as any. No real improvement on gain can be expected by any new tricky design; as far as *F/B* ratio is concerned, it will soon be apparent that one can "tune up" the maximum *F/B* ratio starting with almost any of these schedules.

A summary of raw performance of all of these cases is shown in Table 3.16, where, in addition to data already shown in the previous tables, information on pattern (not explicitly shown here) has been added. This table shows that all cases produce about the same gain; the very small variations are due to the effective "illumination" pattern of the boom aperture. The *F/B* ratio at central gain frequency ($F = 1.0$) varies somewhat, but a very slight change in operating frequency would easily make them all comparable. The frequency position of maximum F/B ratio and the angle of the second null at $F = 1.0$ are related. Lower frequencies of maximum *F/B* should correspond (at central frequency) to longer effective boom illuminated apertures, thus corresponding to lower null angles at $F = 1.0$ and somewhat higher gain. An examination of Table 3.16 shows all of these quantities to be well correlated; it appears therefore that all results are understood and self-consistent.

Table 3.9 — Schedule of director placement on boom for the six six-element Yagi-Uda performance characteristics listed in Table 3.10 through 3.15.

Table	taper type	Refl	element position on boom (λ) DR	D1	D2	D3	D4
3.10	Linear	0	0.200	0.3750	0.5250	0.650	0.750
3.11	Linear	0	0.175	0.3375	0.4875	0.625	0.750
3.12	Linear	0	0.125	0.2625	0.4125	0.575	0.750
3.13	Linear	0	0.100	0.2250	0.3750	0.500	0.750
3.14	Parabolic	0	0.200	0.3167	0.4333	0.550	0.750
3.15	Parabolic	0	0.175	0.3083	0.4417	0.575	0.750

From all of this information it is reasonable to draw a general conclusion that the simplistic Yagi design gives about as much gain as any other design of the same boom length. Tapering element lengths or element position intervals along the boom is of no apparent value. The characteristic of the directors which is important is the average length (or average free space resonant frequency). But this conclusion has been demonstrated only for a boom length of 0.75λ; we must be careful not to generalize too much. Recall that the NBS data (see Fig. 7 of the NBS report) suggested that for booms longer than one wavelength some improvement in gain over simplistic Yagi performance could be obtained with particular director length schedules.[1] My calculations support the NBS result; nevertheless, for boom lengths shorter than one wavelength the simplistic design is as good as any.

[1] Viezbicke, P., "Yagi Antenna Design," *NBS Technical Note 688*, U.S. Department of Commerce, Washington, D.C., December, 1976.

Table 3.10 — Performance characteristics of a six-element Yagi beam with a boom length of 0.75λ, director spacing tapered linearly, large positive interval, directors crowded toward D4.

frequency (F)	*gain (dBi)*	*F/B ratio (dB)*	*feedpoint resistance (ohms)*	*feedpoint reactance (ohms)*
0.970	9.87	11.8	43.7	−38.6
0.972	9.91	12.5	42.7	−36.9
0.974	9.95	13.3	41.6	−35.1
0.976	9.99	14.1	40.4	−33.1
0.978	10.03	15.0	39.1	−31.0
0.980	10.08	16.0	37.7	−28.8
0.982	10.12	17.1	36.3	−26.4
0.984	10.17	18.3	34.9	−23.9
0.986	10.22	19.5	33.5	−21.2
0.988	10.27	20.7	32.0	−18.3
0.990	10.33	21.7	30.6	−15.3
0.992	10.38	22.2	29.3	−12.2
0.994	10.44	22.0	28.0	−9.0
0.996	10.49	21.0	26.7	−5.6
0.998	10.54	19.7	25.5	−2.1
1.000	10.59	18.3	24.4	1.6
1.002	10.63	16.9	23.4	5.3
1.004	10.66	15.6	22.5	9.2
1.006	10.68	14.4	21.7	13.2
1.008	10.69	13.3	21.1	17.2
1.010	10.68	12.3	20.5	21.4
1.012	10.65	11.3	20.1	25.7
1.014	10.60	10.4	19.9	30.0
1.016	10.53	9.6	19.8	34.4
1.018	10.43	8.8	19.9	38.9
1.020	10.30	8.1	20.2	43.5
1.022	10.14	7.4	20.7	48.2
1.024	9.96	6.8	21.4	52.9
1.026	9.76	6.2	22.5	57.8
1.028	9.53	5.7	23.8	62.6
1.030	9.29	5.2	25.6	67.6

Table 3.11 — Performance characteristics of a six-element Yagi beam with a boom length of 0.75λ, director spacing tapered linearly as shown in Table 3.9 (mild positive linear interval).

frequency (F)	*gain (dBi)*	*F/B ratio (dB)*	*feedpoint resistance (ohms)*	*feedpoint reactance (ohms)*
0.970	10.04	12.8	33.0	−37.5
0.972	10.10	13.7	32.5	−35.3
0.974	10.15	14.7	31.8	−32.9
0.976	10.20	15.8	31.1	−30.5
0.978	10.26	17.1	30.4	−28.0
0.980	10.31	18.5	29.6	−25.4
0.982	10.36	20.2	28.8	−22.8
0.984	10.41	22.1	27.9	−20.0
0.986	10.46	24.2	27.1	−17.1
0.988	10.51	26.3	26.3	−14.2
0.990	10.56	27.2	25.5	−11.1
0.992	10.60	26.0	24.7	−8.0
0.994	10.65	23.7	23.9	−4.7
0.996	10.69	21.5	23.2	−1.4
0.998	10.72	19.6	22.6	2.1
1.000	10.75	17.9	22.0	5.6
1.002	10.77	16.4	21.5	9.3
1.004	10.78	15.1	21.1	13.0
1.006	10.78	14.0	20.8	16.8
1.008	10.76	12.9	20.6	20.7
1.010	10.73	12.0	20.6	24.7
1.012	10.68	11.1	20.7	28.7
1.014	10.62	10.4	20.9	32.9
1.016	10.53	9.6	21.3	37.1
1.018	10.43	9.0	21.9	41.5
1.020	10.31	8.3	22.8	45.9
1.022	10.17	7.8	23.9	50.4
1.024	10.02	7.3	25.4	54.9
1.026	9.86	6.8	27.3	59.6
1.028	9.68	6.4	29.8	64.3
1.030	9.51	6.0	32.9	69.0

Table 3.12 — Performance characteristics of a six-element Yagi beam with a boom length of 0.75λ, director spacing tapered linearly as shown in Table 3.9 (mild negative linear interval).

frequency (F)	*gain (dBi)*	*F/B ratio (dB)*	*feedpoint resistance (ohms)*	*feedpoint reactance (ohms)*
0.970	9.88	11.7	17.8	−41.2
0.972	10.02	13.0	17.8	−38.1
0.974	10.16	14.3	17.8	−35.0
0.976	10.27	15.9	17.8	−31.9
0.978	10.38	17.7	17.8	−28.7
0.980	10.47	19.8	17.8	−25.6
0.982	10.55	22.6	17.8	−22.5
0.984	10.63	26.3	17.8	−19.3
0.986	10.69	31.9	17.8	−16.1
0.988	10.75	34.6	17.8	−12.9
0.990	10.81	28.6	17.8	−9.6
0.992	10.85	24.4	17.9	−6.3
0.994	10.89	21.5	18.0	−2.9
0.996	10.91	19.3	18.1	0.5
0.998	10.93	17.6	18.3	4.0
1.000	10.94	16.2	18.6	7.5
1.002	10.94	15.0	19.0	11.1
1.004	10.93	13.9	19.5	14.8
1.006	10.91	13.0	20.1	18.5
1.008	10.88	12.2	20.9	22.3
1.010	10.84	11.4	21.9	26.2
1.012	10.79	10.8	23.2	30.1
1.014	10.73	10.2	24.7	34.1
1.016	10.66	9.7	26.6	38.1
1.018	10.59	9.2	28.9	42.2
1.020	10.51	8.8	31.9	46.1
1.022	10.42	8.5	35.6	49.9
1.024	10.34	8.2	40.2	53.3
1.026	10.25	8.0	46.0	55.8
1.028	10.17	7.9	53.2	56.9
1.030	10.10	7.8	61.5	55.3

Table 3.13 — Performance characteristics of a six-element Yagi beam with a boom length of 0.75λ, director spacing tapered linearly with the elements crowded toward the reflector (large negative interval taper).

frequency *(F)*	*gain* *(dBi)*	*F/B ratio* *(dB)*	*feedpoint resistance* *(ohms)*	*feedpoint reactance* *(ohms)*
0.970	9.27	8.8	12.0	−47.1
0.972	9.57	10.2	12.1	−43.5
0.974	9.82	11.6	12.2	−40.0
0.976	10.04	13.2	12.3	−36.5
0.978	10.23	15.0	12.5	−33.0
0.980	10.38	17.0	12.7	−29.6
0.982	10.52	19.4	12.9	−26.1
0.984	10.63	22.2	13.2	−22.7
0.986	10.73	25.2	13.5	−19.2
0.988	10.81	26.6	13.8	−15.7
0.990	10.87	24.9	14.1	−12.3
0.992	10.93	22.3	14.5	−8.8
0.994	10.96	20.0	15.0	−5.2
0.996	10.99	18.1	15.5	−1.7
0.998	11.01	16.6	16.1	1.9
1.000	11.01	15.3	16.8	5.6
1.002	11.01	14.2	17.6	9.2
1.004	11.00	13.3	18.6	13.0
1.006	10.97	12.5	19.8	16.7
1.008	10.94	11.7	21.2	20.5
1.010	10.90	11.1	22.9	24.4
1.012	10.85	10.6	24.9	28.2
1.014	10.80	10.1	27.4	31.9
1.016	10.74	9.7	30.5	35.6
1.018	10.69	9.3	34.3	38.9
1.020	10.63	9.0	38.9	41.7
1.022	10.57	8.8	44.6	43.5
1.024	10.52	8.7	51.2	43.6
1.026	10.47	8.6	58.4	41.0
1.028	10.44	8.6	64.6	34.7
1.030	10.41	8.8	67.3	24.8

Table 3.14 — Performance characteristics of a six-element Yagi beam with a boom length of 0.75λ, director spacing tapered pseudo parabolically according to the schedule of Table 3.9 (large positive interval).

frequency (F)	*gain (dBi)*	*F/B ratio (dB)*	*feedpoint resistance (ohms)*	*feedpoint reactance (ohms)*
0.970	10.25	15.5	33.7	−34.3
0.972	10.30	16.5	33.4	−31.8
0.974	10.35	17.6	33.0	−29.2
0.976	10.40	18.9	32.6	−26.5
0.978	10.44	20.4	32.2	−23.9
0.980	10.49	22.1	31.8	−21.0
0.982	10.53	24.1	31.3	−18.2
0.984	10.56	26.3	30.9	−15.3
0.986	10.60	28.2	30.5	−12.3
0.988	10.63	28.6	30.1	−9.2
0.990	10.66	27.0	29.7	−6.0
0.992	10.69	24.8	29.4	−2.8
0.994	10.71	22.6	29.1	0.6
0.996	10.73	20.7	28.9	4.0
0.998	11.74	19.1	28.7	7.6
1.000	11.75	17.6	28.7	11.3
1.002	11.75	16.4	28.7	15.0
1.004	10.74	15.3	28.9	18.9
1.006	10.72	14.2	29.2	22.9
1.008	10.69	13.3	29.7	27.1
1.010	10.63	12.5	30.5	31.3
1.012	10.59	11.7	31.5	35.7
1.014	10.52	10.9	32.8	40.3
1.016	10.44	10.3	34.5	45.0
1.018	10.35	9.6	36.7	49.9
1.020	10.25	9.0	39.5	54.9
1.022	10.13	8.5	43.2	60.1
1.024	10.01	8.0	47.9	65.2
1.026	9.88	7.5	54.0	70.3
1.028	9.75	7.1	62.1	74.8
1.030	9.61	6.7	72.8	78.2

Table 3.15 — Performance characteristics of a six-element Yagi beam with a boom length of 0.75λ, director spacing tapered pseudo parabolically according to the schedule of Table 3.9 (mild positive interval).

frequency (F)	*gain (dBi)*	*F/B ratio (dB)*	*feedpoint resistance (ohms)*	*feedpoint reactance (ohms)*
0.970	10.20	14.0	31.0	−35.2
0.972	10.26	15.0	30.7	−32.7
0.974	10.32	16.1	30.4	−30.2
0.976	10.37	17.4	30.1	−27.5
0.978	10.42	18.9	29.7	−24.9
0.980	10.47	20.6	29.3	−22.2
0.982	10.52	22.8	28.9	−19.4
0.984	10.56	25.6	28.5	−16.6
0.986	10.60	29.6	28.1	−13.6
0.988	10.64	35.4	27.7	−10.6
0.990	10.67	34.8	27.3	−7.6
0.992	10.70	29.1	27.0	−4.4
0.994	10.73	25.2	26.6	−1.2
0.996	10.76	22.5	26.4	2.1
0.998	10.77	20.3	26.2	5.6
1.000	10.78	18.6	26.1	9.1
1.002	10.79	17.1	26.1	12.7
1.004	10.78	15.9	26.2	16.4
1.006	10.77	14.8	26.4	20.2
1.008	10.74	13.8	26.8	24.0
1.010	10.70	12.9	27.4	28.0
1.012	10.66	12.1	28.2	32.1
1.014	10.60	11.3	29.3	36.3
1.016	10.52	10.7	30.7	40.6
1.018	10.44	10.0	32.8	45.0
1.020	10.34	9.5	34.8	49.4
1.022	10.24	9.0	37.7	53.9
1.024	10.12	8.5	41.5	58.2
1.026	10.00	8.1	46.3	62.4
1.028	9.88	7.7	52.5	66.0
1.030	9.76	7.5	60.4	68.4

Front-to-Back Optimization

It can be seen from Table 3.16 that the best "null" positions give quite different values of maximum *F/B*. Indeed, good nulls or correspondingly high values of *F/B* must be viewed as accidental vectorial cancellations in the reverse or back direction; such good cancellations will not generally occur with any arbitrary boom illumination. Note that the various cases shown in Table 3.16 display maximum *F/B* ratios ranging from 22 to 40 dB. It is an interesting exercise to see if there is some way to significantly enhance the maximum *F/B* ratio by some variational procedure.

Let us start with the simplistic Yagi design (Fig. 3.1) and vary the position of, say D3, along the boom. We now know that small variations in position will not significantly affect gain, but vectorial cancellation effects in the back direction can be expected to be significant. If we find a position for D3 which maximizes the *F/B* ratio, its vectorial contribution in the back direction should be approximately out of phase with the residue from all other elements. At this point some other element (say D1) can be positioned for a new (still higher) maximum *F/B* ratio; after this is done D3 can be readjusted again for a new maximum *F/B* ratio, etc. By iterating the two adjustments it should be possible to continuously improve *F/B* ratios, presumably to as high a value as desired. With such an iteration procedure it is desirable to start with a fairly good value of *F/B* so that only small variations in element position can have significant effect.

Table 3.16 — Performance comparison of six-element Yagi beams with varying director lengths and element positions along the boom shows little gain variation.

from Table	gain (dBi)	F=1 F/B (dB)	maximum F/B (dB)	maximum F	angle of m = 2 null (deg)
3.1	10.86	17.0	38.0	0.988	144
3.3	10.78	18.4	40.0	0.988	144
3.4	10.65	16.5	31.9	0.986	138
3.5	11.03	11.6	29.7	0.980	138
3.6	10.96	14.2	31.6	0.984	141
3.7	10.73	19.9	38.7	0.992	150
3.8	10.58	22.0	33.6	0.994	153
3.10	10.59	18.3	22.2	0.992	150
3.11	10.75	17.9	27.2	0.990	147
3.12	10.94	16.2	34.6	0.988	141
3.13	11.01	15.3	26.6	0.988	138
3.14	10.75	17.7	28.6	0.988	144
3.15	10.78	18.6	35.4	0.988	147

I have carried out such an iteration (using D3 and D1) for the simplistic Yagi design and have arrived at the following positions:

element	*position (x)*
Reflector	0.000λ
Driven Element	0.150λ
Director 1	0.28967λ
Director 2	0.450λ
Director 3	0.58945λ
Director 4	0.750λ

This design, optimized for maximum F/B ratio (at $F = 0.990$) produces the performance displayed in Table 3.17. A careful comparison of this table with the original simplistic model shows virtually identical performance in all respects except that the F/B maximum has gone up from an excellent 38 dB to an astounding 98 dB. Even this high value is not a real limit; it is limited only by the number of iterations which were made.

Notice that this astronomical value of F/B is of no practical significance. It occurs at essentially a single frequency and its effective bandwidth is vanishingly small. Moreover, extremely small variations in the Yagi dimensions will upset the cancellation; in practice one could not likely construct a mechanically satisfactory, fully optimized Yagi antenna. Nevertheless, the mathematical iteration shows that it is possible, in principle, to obtain (at a single frequency) an arbitrarily high F/B ratio. It is likely that there are a large number of potential solutions involving iterations with other elements. Furthermore, we now know that the variations in F/B maxima shown in Table 3.16 result from the particular illumination chosen, and it is very likely that minor element placement variations could make an arbitrarily high F/B ratio design starting from any of the cases shown.

To understand this iteration procedure it is helpful to show the vectorial contributions of each element to the forward and back waves. The (current) contribution from a given element will be a vector whose magnitude is the magnitude of element current and whose phase consists of two parts. The first part is the actual (time) phase of the element current referred to some time origin (say the driver current) and the second is the (space) phase change due to the element position referred to some space origin (say at $x = 0$ along the boom). Note that this second part changes sign in going from a forward wave to a reverse wave. Figure 3.3 shows these (current) vectorial contributions at $F = 0.988$ for the original simplistic Yagi design (Fig. 3.1) to both forward and reverse waves.

Note that for the forward wave the individual element contributions do not all fully reinforce the forward wave; in fact, the contribution from the reflector is even negative with respect to the final total (current) vector. This curious result is typical of all Yagi arrays. Note that the contributions to

Table 3.17 — Performance vs frequency characteristics of a six-element Yagi beam with a boom length of 0.75 λ, element positions optimized for maximum F/B ratio (at *F* = 0.990).

frequency (F)	*gain (dBi)*	*F/B ratio (dB)*	*feedpoint resistance (ohms)*	*feedpoint reactance (ohms)*
0.960	9.48	8.1	26.8	−50.7
0.962	9.63	8.8	26.9	−47.9
0.964	9.76	9.7	27.1	−45.2
0.966	9.88	10.5	27.2	−42.5
0.968	9.98	11.4	27.2	−39.8
0.970	10.07	12.4	27.2	−37.2
0.972	10.15	13.4	27.1	−34.5
0.974	10.23	14.5	27.0	−31.8
0.976	10.30	15.8	26.9	−29.2
0.978	10.36	17.2	26.7	−26.5
0.980	10.42	18.8	26.5	−23.7
0.982	10.48	20.8	26.2	−20.9
0.984	10.53	23.4	26.0	−18.0
0.986	10.58	26.9	25.7	−15.1
0.988	10.62	33.0	25.5	−12.1
0.990	10.66	98.8	25.2	−9.1
0.992	10.70	33.0	25.0	−5.9
0.994	10.73	27.0	24.9	−2.7
0.996	10.76	23.5	24.7	0.6
0.998	10.79	21.0	24.7	3.9
1.000	10.80	19.1	24.7	7.4
1.002	10.81	17.5	24.8	10.9
1.004	10.81	16.2	25.1	14.6
1.006	10.80	15.0	25.5	18.3
1.008	10.78	14.0	26.0	22.1
1.010	10.75	13.1	26.8	26.0
1.012	10.71	12.3	27.8	30.0
1.014	10.65	11.5	29.2	34.0
1.016	10.59	10.9	30.8	38.1
1.018	10.51	10.3	33.0	42.3
1.020	10.42	9.8	35.8	46.4
1.022	10.32	9.3	39.3	50.4
1.024	10.22	8.9	43.7	54.1
1.026	10.11	8.6	49.4	57.1
1.028	10.00	8.3	56.5	58.9
1.030	9.88	8.1	65.1	58.4
1.032	9.78	7.9	74.6	54.2
1.034	9.68	7.9	83.0	44.4
1.036	9.59	8.0	85.9	28.9
1.038	9.51	8.2	79.0	12.2
1.040	9.44	8.6	63.5	1.5

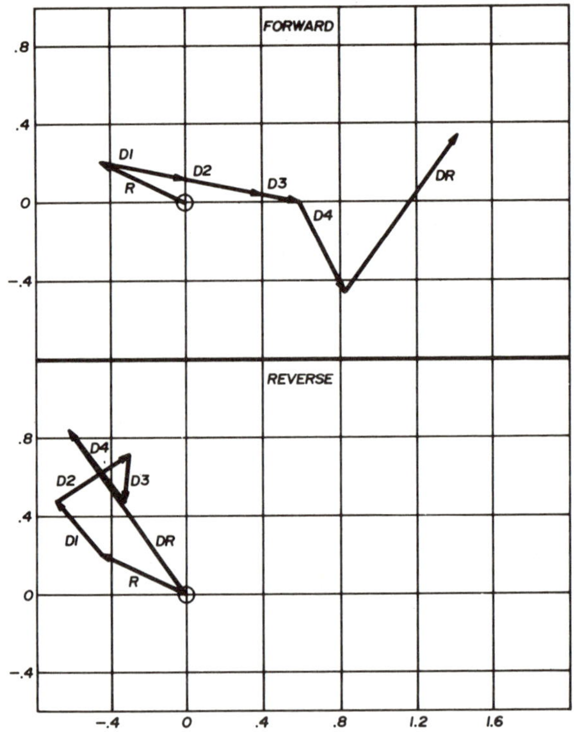

Fig. 3.3 — Current vectorial contributions at F = 0.988 for six-element Yagi (boom = 0.75λ), forward and reverse waves.

the back or reverse wave, in total, nearly cancel out, leaving only a small residue which accounts for the 38 dB *F/B* ratio.

Now it is easy to see conceptually what happens in the iterative procedure to reduce the reverse wave residual. If you look at the reverse wave vector plot, it is easy to imagine that as D3 is moved along the boom, the D3 vector rotates around its origin. The reverse wave residual will then be changed along an axis at right angles to the D3 vector and can be minimized by the D3 position. After this is done another element, say D1, can be moved along the boom; its vector contribution is at a different angle and can therefore reduce the residual still further. Thus, in principle, iterative motions of two elements whose reverse wave vectors contribute at different angles can ultimately reduce the reverse wave residual to as low a value as desired. The iterative convergence will be most rapid if the two element vectors are orthogonal; nevertheless, it can converge adequately for many element combinations.

Of course, this conceptual picture is oversimplified; as any element is moved

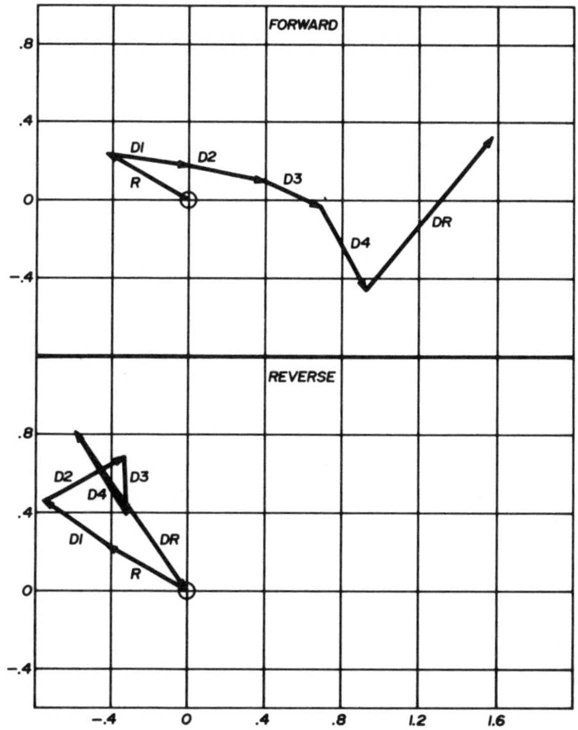

Fig. 3.4 — Current vectorial contributions for the optimized six-element Yagi at F = 0.990.

on the boom not only does its vector rotate, but all element currents and phases readjust somewhat. However, these readjustments are usually minor and in practice cause little difficulty as long as you start with a reasonably small residual as shown in Fig. 3.3. Figure 3.4 shows the vectorial contributions for the optimized Yagi beam. Note that the element contributions are only slightly modified in the optimization procedure.

At this point I must issue a warning. Recall that the mathematical model being used in these computations involves certain approximations. These approximations make relatively little difference in the calculations for forward gain, but they become crucial in calculations involving vectorial cancellation or closure for back radiation. Thus the explicitly calculated positions for and magnitude of a very high F/B ratio are not to be trusted. Nevertheless, the general behavior is still valid. The real Yagi can still be made to have a high F/B ratio, just as our mathematical model shows, but it may occur at a slightly different frequency and it may require slightly different positions for D3 and D1 in the final optimization.

Optimum Design

We now have the necessary tools with which to design truly excellent Yagi antennas. We start first from a knowledge that the boom length should be approximately an odd number of quarter-wavelengths for an initial simplistic design; we have seen that such a boom length promotes an inherently high *F/B* ratio at a frequency near the center of the best gain band; boom length also determines ultimate gain. After the boom length is chosen a resonant frequency schedule is chosen (see appropriate figures from Chapter 2) for reflector and director(s). A preliminary calculation is then made to accurately determine the frequency of maximum *F/B* ratio, which will not necessarily correspond with the frequency at the center of the best gain portion.

The useful band is now to be centered around the *F/B* point; it is necessary to ensure that there is enough gain bandwidth left for the intended purpose. Remember that the overall gain bandwidth is basically controlled by the resonant frequency schedule of the parasites. This bandwidth should not be larger than necessary, because gain is compromised somewhat as the bandwidth increases.

Now translate the frequency (F_1) of best *F/B* to $F = 1.0$ by multiplying all parasite lengths by F_1; a new preliminary calculation, possibly iterated once more, will insure that the best *F/B* ratio is exactly $F = 1.0$. Next, alternately vary the *x* position of D3 and then D1 to get larger and larger values of *F/B* at $F = 1.0$ until the value is sufficiently high.

Design Example

An example illustrates this design procedure. Let's choose a boom length of 0.780λ. From an inspection of the results of our test Yagi simplistic design (Fig. 3.1) we can probably use the same parasite resonant frequency schedule and still obtain an adequate ultimate gain bandwidth performance. Listed below are the initial element positions along the boom:

element	*position x (λ)*	*initial length (λ)*	*intermediate length (λ)*
Reflector	0.000	0.50195	0.49343
Driven El	0.156	0.48167	0.48167
Director 1	0.312	0.45414	0.44643
Director 2	0.468	0.45414	0.44643
Director 3	0.624	0.45414	0.44643
Director 4	0.780	0.45414	0.44643

Initial performance of this Yagi model is shown in Table 3.18; the frequency for maximum *F/B* is $F = 0.984$. Shortening all elements by approximately this frequency factor yields the intermediate design also shown above.

Table 3.18 — Performance characteristics of the six-element Yagi discussed in the text (boom length = 0.78λ).

frequency (F)	*gain (dBi)*	*F/B ratio (dB)*	*feedpoint resistance (ohms)*	*feedpoint reactance (ohms)*
0.970	10.20	14.8	23.6	−35.8
0.972	10.29	16.2	23.3	−32.9
0.974	10.38	17.7	23.1	−29.9
0.976	10.46	19.6	22.9	−26.9
0.978	10.54	21.8	22.6	−23.9
0.980	10.61	24.6	22.4	−20.8
0.982	10.67	28.3	22.1	−17.7
0.984	10.73	31.8	21.9	−14.6
0.986	10.79	30.2	21.7	−11.4
0.988	10.84	26.3	21.5	−8.1
0.990	10.88	23.2	21.4	−4.8
0.992	10.92	20.8	21.3	−1.4
0.994	10.95	18.9	21.2	2.1
0.996	10.97	17.4	21.2	5.6
0.998	10.98	16.0	21.3	9.2
1.000	10.98	14.9	21.5	12.8
1.002	10.98	13.9	21.8	16.5
1.004	10.96	13.0	22.2	20.3
1.006	10.92	12.1	22.8	24.2
1.008	10.88	11.4	23.5	28.1
1.010	10.83	10.8	24.4	32.1
1.012	10.76	10.2	25.6	36.1
1.014	10.68	9.6	27.0	40.2
1.016	10.59	9.1	28.8	44.3
1.018	10.49	8.7	31.0	48.5
1.020	10.39	8.3	33.7	52.5
1.022	10.28	8.0	37.2	56.5
1.024	10.17	7.7	41.4	60.1
1.026	10.06	7.4	46.6	63.2
1.028	9.95	7.3	53.1	65.3
1.030	9.85	7.2	60.9	65.5

Performance for this intermediate design is shown in Table 3.19 Note that since all lengths were not scaled (boom not scaled), this intermediate Yagi is not really quite the same as our starting model; the maximum *F/B* ratio has, in fact, fallen to 27 dB. However, this is of no concern; it is now time to iteratively vary D3 and D1 positions to "tune up" the *F/B* ratio. Alternatively, if our concept of optimization is correct, iterative variations of D3

Table 3.19 — Performance characteristics of the intermediate design six-element Yagi described in the text; maximum *F/B* at *F* = 1.0.

frequency (F)	*gain (dBi)*	*F/B ratio (dB)*	*feedpoint resistance (ohms)*	*feedpoint reactance (ohms)*
0.970	9.08	7.6	24.3	−40.5
0.972	9.27	8.4	24.2	−37.5
0.974	9.45	9.2	24.0	−34.5
0.976	9.61	10.0	23.9	−31.6
0.978	9.76	10.9	23.8	−28.6
0.980	9.89	11.8	23.7	−25.7
0.982	10.02	12.8	23.5	−22.8
0.984	10.13	13.9	23.3	−19.9
0.986	10.24	15.1	23.2	−16.9
0.988	10.33	16.5	23.0	−13.9
0.990	10.42	18.0	22.8	−11.0
0.992	10.51	19.7	22.6	−8.0
0.994	10.59	21.7	22.4	−4.9
0.996	10.66	24.0	22.2	−1.8
0.998	10.73	26.2	22.0	1.3
1.000	10.79	27.1	21.8	4.5
1.002	10.85	25.9	21.7	7.7
1.004	10.89	23.8	21.6	10.9
1.006	10.94	21.7	21.5	14.3
1.008	10.97	19.8	21.5	17.6
1.010	11.00	18.2	21.6	21.1
1.012	11.02	16.9	21.7	24.5
1.014	11.03	15.7	21.8	28.1
1.016	11.03	14.6	22.1	31.7
1.018	11.03	13.7	22.5	35.4
1.020	11.01	12.9	23.0	39.1
1.022	10.97	12.1	23.7	42.9
1.024	10.93	11.4	24.5	46.7
1.026	10.88	10.8	25.5	50.6
1.028	10.81	10.2	26.8	54.6
1.030	10.74	9.7	28.3	58.5

and DR could also tune up the *F/B* ratio. I have carried out both iterations and the resulting optimized Yagi parameters are as shown:

element	*length* (λ)	*D3-D1 opt* x (λ)	*D3-DR opt* x (λ)
Reflector	0.49343	0.000	0.000
Driven el	0.48167	0.156	0.175595
Director 1	0.44643	0.291564	0.312
Director 2	0.44643	0.468	0.468
Director 3	0.44643	0.64075	0.6328873
Director 4	0.44643	0.780	0.780

Table 3.20 — Performance of the six-element Yagi where the positions of directors D1 and D3 have been varied to "tune up" the *F/B* ratio.

frequency (F)	*gain (dBi)*	*F/B ratio (dB)*	*feedpoint resistance (ohms)*	*feedpoint reactance (ohms)*
0.970	9.06	8.0	27.2	−43.1
0.972	9.24	8.7	27.1	−40.3
0.974	9.39	9.5	27.0	−37.5
0.976	9.53	10.4	26.9	−34.7
0.978	9.66	11.3	26.7	−31.9
0.980	9.78	12.2	26.5	−29.2
0.982	9.89	13.2	26.3	−26.4
0.984	9.99	14.4	26.1	−23.6
0.986	10.09	15.6	25.8	−20.8
0.988	10.18	17.1	25.5	−17.9
0.990	10.26	18.7	25.3	−15.1
0.992	10.34	20.7	25.0	−12.2
0.994	10.42	23.3	24.7	−9.2
0.996	10.49	26.9	24.4	−6.2
0.998	10.56	32.9	24.1	−3.1
1.000	10.63	119.8	23.8	0.0
1.002	10.69	33.0	23.6	3.2
1.004	10.75	27.0	23.4	6.4
1.006	10.80	23.5	23.2	9.7
1.008	10.85	21.0	23.1	13.1
1.010	10.90	19.1	23.1	16.6
1.012	10.94	17.6	23.1	20.1
1.014	10.99	16.2	23.2	23.7
1.016	10.99	15.1	23.4	27.4
1.018	11.00	14.1	23.7	31.1
1.020	11.01	13.2	24.2	35.0
1.022	11.00	12.4	24.8	38.9
1.024	10.97	11.7	25.6	42.9
1.026	10.96	11.0	26.6	47.0
1.028	10.92	10.4	28.0	51.2
1.030	10.87	9.9	29.6	55.4

Performance of the D3-D1 optimized antenna is shown in Table 3.20; it is nearly the same as that of the intermediate design (Table 3.19) except that the F/B ratio at $F = 1.0$ has gone up from 27 dB to an astronomical 120 dB. Similarly, the performance for the D3-DR optimized antenna is shown in Table 3.21. Again an astounding F/B ratio figure is achieved; moreover, the newer optimized beam performance is essentially identical with that of the first optimized model.

It is instructive to examine the final vector contributions to forward and reverse waves; Fig. 3.5 shows such a plot for the D3-D1 optimized Yagi and Fig. 3.6 a similar plot for the D3-DR optimized model. Note that they look

Table 3.21 — Performance of the six-element Yagi where the positions of the driven element (DR) and director (D3) have been optimized through computer iteration.

frequency (F)	*gain (dBi)*	*F/B ratio (dB)*	*feedpoint resistance (ohms)*	*feedpoint reactance (ohms)*
0.970	9.43	9.2	31.6	−40.7
0.972	9.55	9.9	31.5	−38.1
0.974	9.67	10.7	31.3	−35.5
0.976	9.77	11.5	31.0	−33.0
0.978	9.87	12.3	30.8	−30.4
0.980	9.96	13.2	30.4	−27.8
0.982	10.04	14.2	30.1	−25.2
0.984	10.12	15.3	29.7	−22.6
0.986	10.19	16.5	29.3	−19.9
0.988	10.27	17.9	28.9	−17.2
0.990	10.33	19.6	28.4	−14.5
0.992	10.40	21.5	28.0	−11.7
0.994	10.47	24.1	27.5	−8.8
0.996	10.53	27.6	27.0	−5.8
0.998	10.59	33.7	26.6	−2.9
1.000	10.65	150.3	26.2	0.2
1.002	10.70	33.7	25.8	3.3
1.004	10.75	27.7	25.4	6.5
1.006	10.80	24.1	25.0	9.8
1.008	10.85	21.6	24.8	13.1
1.010	10.89	19.7	24.5	16.5
1.012	10.92	18.1	24.4	20.0
1.014	10.95	16.7	24.3	23.6
1.016	10.97	15.6	24.3	27.2
1.018	10.98	14.5	24.4	30.9
1.020	10.98	13.6	24.7	34.7
1.022	10.98	12.8	25.1	38.6
1.024	10.96	12.0	25.6	42.5
1.026	10.93	11.3	26.4	46.5
1.028	10.89	10.7	27.3	50.7
1.030	10.83	10.1	28.5	54.8

similar, differing only in minute details. Incidentally, it is noteworthy that the reverse plots show vectorial contributions going around the clock twice, corresponding to the second null, which we have constructed.

There is one final point worth mentioning. An examination of Tables 3.1, 3.20, and 3.21 reveals that the frequency for the best *F/B* ratio is not generally quite the same as the frequency center of the gain bandwidth. It is offset by an amount that depends only on the boom length. This offset is of small importance as long as the gain bandwidth is large enough; it is nevertheless possible to empirically measure the offset frequency as a function of boom

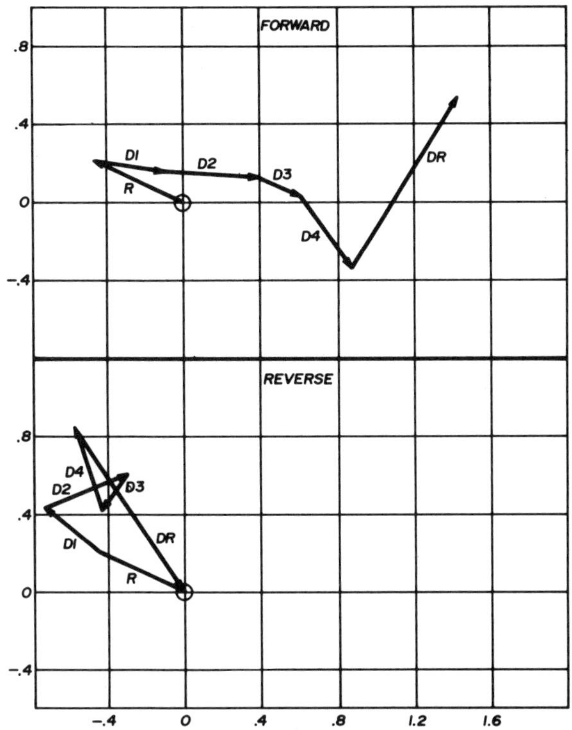

Fig. 3.5 — Current vectorial contributions of the D3-D1 optimized Yagi.

length. Let us fix the frequency of best *F/B* ratio as $F = 1.0$, and designate the frequency of (central) best gain (4 percent BW) as F_G (offset frequency $= F_G - 1.0$); empirical results are shown in Fig. 3.7. Note that if the boom length is 0.63λ the offset disappears. For booms shorter than this value the offset is negative, and for booms longer than 0.63λ the offset is positive. But it is clearly possible to design a satisfactory Yagi over a considerable range of boom lengths without incurring an offset which is comparable to the bandwidth itself; it is only necessary to take this offset into account in fixing the original bandwidth over which gain must be high. From a gain consideration alone the longer booms are best; that is why the example I used for illustrative purposes had a boom length of 0.78λ.

Number of Reflectors

It is interesting to consider a major change in possible Yagi antenna design: to explore the effect of changing the number of reflectors in a Yagi array. Up to this point we have assumed only a single reflector with a variable number of directors. It is tempting to consider increasing the number of reflec-

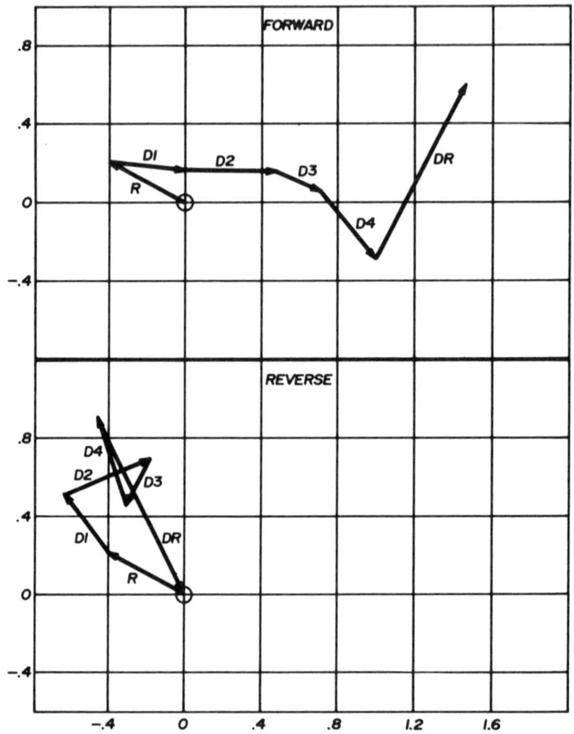

Fig. 3.6 — Current vectorial contributions for the D3-DR optimized Yagi beam.

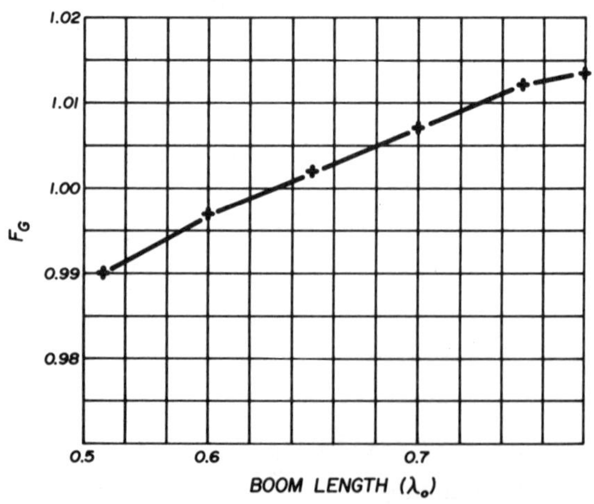

Fig. 3.7 — Plot illustrating frequency ratio for best central gain to best *F/B*. Note that frequency offset disappears for a boom length of 0.63λ.

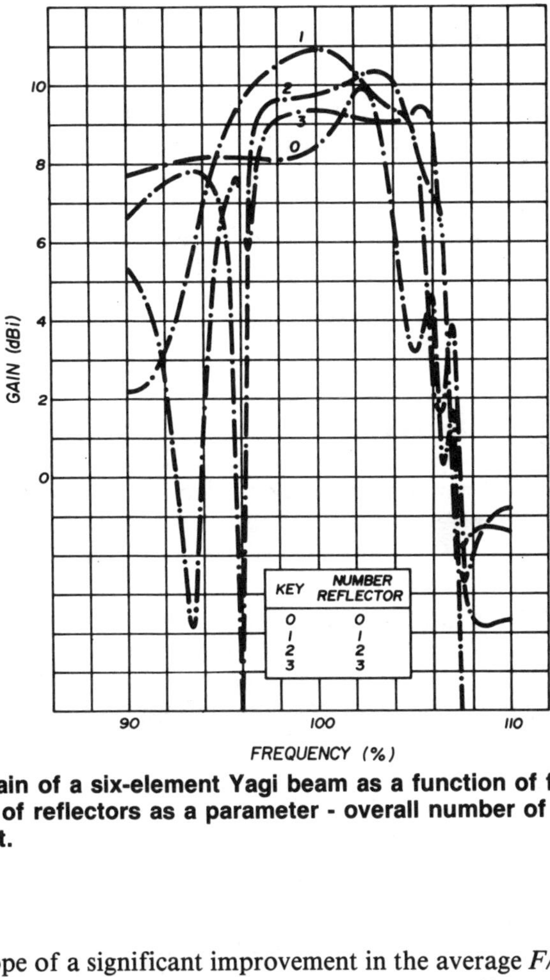

Fig. 3.8 — Gain of a six-element Yagi beam as a function of frequency with number of reflectors as a parameter - overall number of elements held constant.

tors in the hope of a significant improvement in the average *F/B* ratio over the entire bandwidth to be used. This question is now easily explored. I shall assume that the simplistic test Yagi of Fig. 3.1 will be our standard. To keep conditions other than the number of reflectors as constant as possible I shall keep the total boom length constant at 0.75λ, and the total number of parasites constant at five. We shall compare the cases where the number of reflectors is zero, one (our test standard), two, and three. Fig. 3.8 shows frequency-swept gain curves for all four cases; the curves are keyed to the legend on the diagram. Severe resonance effects are noticed near the free space resonances of the reflector ($F = 0.96$) and the directors ($F = 1.06$); these resonances, however, were purposely spread far enough to allow the 4 percent band of interest a good gain figure.

The highest curve (curve 1) displays gain for the standard simplistic Yagi (same as Fig. 3.1) and it is clearly the best performer. The zero reflector case (curve 0) yields substantially less gain in the region of interest; it also contains no resonance effect at the reflector frequency, because there is no reflector. The two- and three-reflector cases (curves 2 and 3) show progressive loss of gain over the original standard; the reason is to be found in the much lower currents induced in the additional reflectors. Shown below are the reflector currents when the driver is excited by one ampere at the central frequency ($F = 1.0$):

number of reflectors	*magnitude of reflector current (amps)*		
1	0.477		
2	0.043	0.538	
3	0.031	0.046	0.626

Note that the reflector next to the driver has substantial current while all other reflectors are hardly excited at all. Thus where there are multiple reflectors, the effective boom length is shortened and we therefore should expect the gain to fall appreciably.

Fig. 3.9 shows the *F/B* ratio for these same four cases. Clearly the standard Yagi antenna (curve 1) is superior to the zero reflector case (curve 0). In the two-reflector case (curve 2) the peak of maximum *F/B* (corresponding to the second null) has moved significantly higher in frequency. We have already learned that this occurs when the effective boom length is reduced (in this case by the relatively ineffective first reflector). This effect is exag-

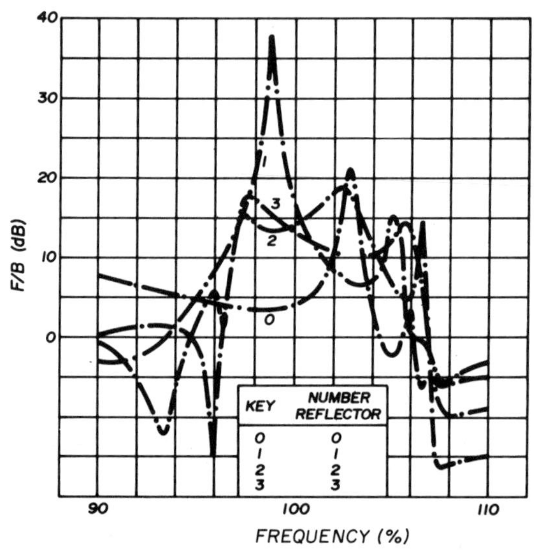

Fig. 3.9 — *F/B* of a six-element Yagi beam as a function of frequency with number of reflectors as a parameter.

gerated in the three-reflector case (curve 3) where the effective boom is still shorter due to the first two relatively ineffective reflectors.

Thus we now see that there is a very good reason why a Yagi should contain one and only one reflector in the linear boom array; one is definitely needed to improve the gain and *F/B*. More than one reflector reduces the effective boom length and therefore gain; also, because of the relatively small currents induced in the extra reflectors, they do very little to the basic Yagi *F/B* ratio potential.

Missing Parasites

A common observation among Amateurs who have had large Yagi antennas in operation over a period of time is that when a parasitic element is broken or even entirely missing the Yagi continues to perform surprisingly well. We may now examine quantitatively just what occurs; for comparison I shall use the same six-element simplistic Yagi design of Fig. 3.1.

When a parasite is missing, the individual element currents all readjust to new values; such a readjustment changes the effective boom illumination function and therefore must cause a change in Yagi antenna performance. Starting with the standard six-element simplistic design, I have made calculations of performance when one parasite is missing. Frequency-swept plots of gain and *F/B* ratio are shown in Figs. 3.10 and 3.11; the individual curves are keyed to the legend in the diagram.

Note that there is significant gain displayed for all of the cases. The greatest

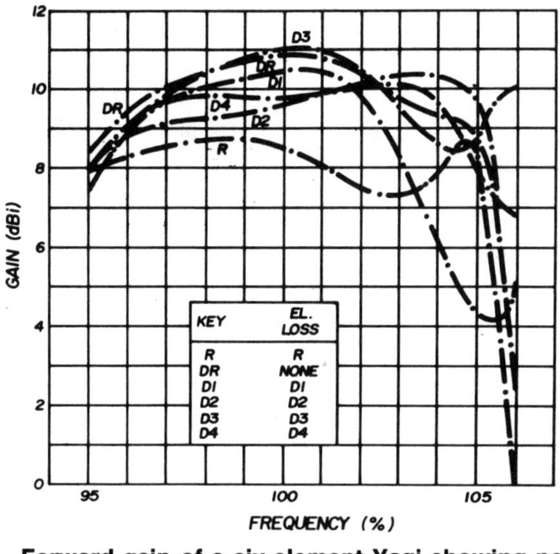

Fig. 3.10 — Forward gain of a six-element Yagi showing performance when one element is missing. *F/B* under similar conditions is plotted in Fig. 3.11.

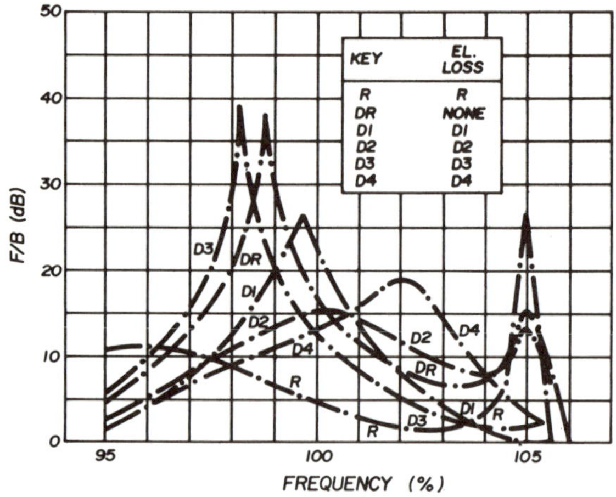

Fig. 3.11 — Front-to-back ratio of a six-element Yagi showing the effect of a missing element. Forward gain under similar conditions is shown in Fig. 3.10.

loss in performance occurs when the reflector, R, is missing; this, of course, is analogous to the previously discussed zero reflector case but now with a shorter (residual) effective boom. The most surprising aspect of Fig. 3.10 is the small but real increase in gain occasioned by the loss of D3. This can only be understood if the readjustment element currents constitute an effective boom illumination function slightly longer than that for the fully populated beam; in this event we would expect the frequency for maximum *F/B* to be lower than that for the standard case. Fig. 3.11 shows this to be true. For all other cases of missing parasites the frequency of maximum *F/B* is increased, indicating a shortened effective boom length and hence lowered gain. The lowered gain is verified in Fig. 3.10.

Thus a missing parasite is not always disastrous. However, if you look at the performance at the frequency of best *F/B*, the original fully populated Yagi is best.

It is now apparent that a Yagi antenna really "wants to work." Even major changes, such as a missing inner director, work surprisingly well due to automatically readjusted element currents. It is now perfectly obvious why the Yagi antenna is so popular: it will provide reasonable performance no matter how it is constructed. It will provide top performance, especially in the *F/B* ratio, only if carefully made in accordance with the design rules presented in this chapter.

Summary

In this chapter, I have explored the effects of departures from the simplistic design previously given. The results show:

1) Director length taper schedules have no apparent beneficial effect on gain or *F/B* for boom lengths smaller than one wavelength. The important design parameter is the average director length — not the taper schedule.

2) Element placement schedules on the boom also have a marginal effect on gain or *F/B* for boom lengths less than one wavelength.

3) The simplistic design is as good as any design for boom lengths less than one wavelength.

4) A Yagi linear array on a given boom is best when it involves one and only one reflector element.

5) The *F/B* ratio at a given design frequency can, in principle, be increased without limit by an iterative design procedure.

6) Very high values of *F/B* will be available only over very narrow bandwidths.

7) The Yagi antenna is basically very tolerant of major faults. Even missing parasitic elements cause surprisingly little deterioration in gain.

CHAPTER 4

LOOP ANTENNAS

Up to this point we have considered only linear cylindrical elements about one-half wavelength long with a small diameter compared to wavelength; there is in common use, however, a radiation "element" consisting of a loop of wire about one wavelength in circumference. There are many such loop configurations: a triangular loop (commonly known as a delta loop), a square loop with two sides parallel to the earth (known as the quad), and a square loop oriented so that two diagonal corners are perpendicular to the earth (known as a diamond loop). Loops can also be made that are not equilateral. Triangles can be isosceles or even with three different sides; four-sided loops can be rectangles, and, indeed loops can have more than four sides or can even be round.

The one-wavelength loop can be used either as a driven radiator or a parasite. To drive the loop it is opened at some point on its circumference where it is excited with a suitable voltage or current. As a parasite the loop is left closed; current will flow depending upon the induced voltages from other loops or elements and the self-impedance of the loop. To help understand the behavior of these loops it is useful to model the loop in terms of dipole elements.

Square Loop Model

Let's first consider a model for a square driven loop. The model starts with two driven half-wavelength dipoles in free space as shown in Fig. 4.1.

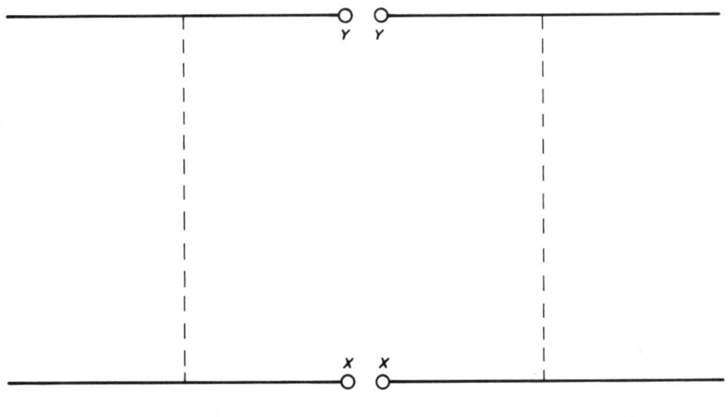

Fig. 4.1 — Diagram of two driven dipoles with one-quarter wavelength vertical spacing. Bending the ends of the dipoles together, as shown by the dotted lines, forms the square loop.

The dipoles are voltage driven at their centers (at *xx* and *yy*) and in this model are separated (vertically) by one-quarter wavelength. If the excitation is equal at *xx* and *yy* (the same voltage and phase), the dipole currents will also be equal and therefore, the voltages at the ends of the dipoles will be equal. The outer eighth-wavelength sections of each dipole may now be bent as shown by the dotted line, forming a square. Since the voltages at the dipole ends are still equal, the bent dipoles can be connected together without changing any currents in the square.

Since for each dipole there is a unique relationship between voltage at its ends and current at its center, the end connection of both dipoles in the square configuration insures that the center dipole currents will be equal even though excitation voltages at *xx* and *yy* may be different. Indeed, we can remove voltage excitation at *yy* (shorting the terminals) and still be sure that whatever dipole currents flow from excitation at *xx*, they will be equal in the two dipoles. One can look at the excitation in this case as a dipole current fed at *xx*; the dipole centered at *yy* is voltage fed at its ends. Note that to realize the same loop current the sum of the voltages supplied to *xx* and *yy* must be constant. This leads to the well-known fact that the driving-point impedance at *xx* with *yy* shorted will be just twice the impedance at *xx* when *yy* and *xx* are excited equally.

Fig. 4.2 shows the square loop excited at the center of the bottom section and also shows the relative magnitudes and directions of the currents which flow. Note that the horizontal sections of the square show currents in the same direction as those of the original dipoles of Fig. 4.1; therefore these sections will provide radiation fields at long distances. Since they are shorter than the original half-wavelength dipoles, however, they require more current to produce a given radiation field, i.e., the radiation resistance will drop. Moreover, the directivity and gain of each segment will be somewhat smaller than that of a full half-wavelength dipole. Both of these effects will be discussed shortly.

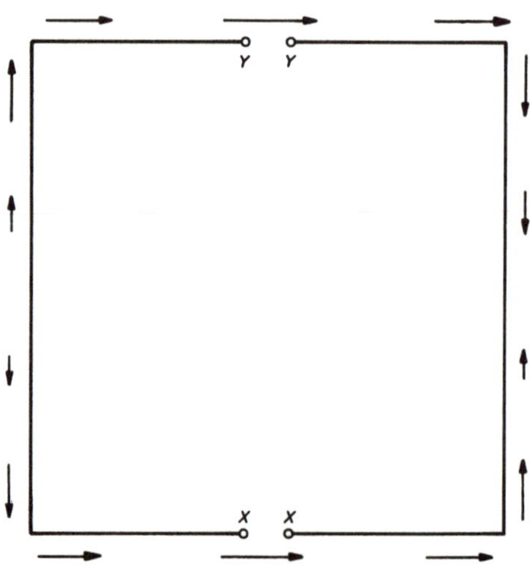

Fig. 4.2 — Outline of the square loop showing the current distribution in the horizontal and vertical members. Horizontal polarization results with the feed points as shown.

The vertical portions of the square loop shown in Fig. 4.2 are quite different. Note that for each of these vertical segments the top and bottom portions have identical currents but in opposite directions. This symmetry assures complete cancellation of the far (radiation) field. In other words, the vertical sections do not radiate; they simply act as a capacitive "top hat" loading for the horizontal radiating segments. Thus the radiating properties of the square loop of Fig. 4.2 will be identical to that of the truncated horizontal segments alone. These will, of course, produce only horizontally polarized radiation.

Table 4.1 — Broadside gain versus separation for two parallel half-wavelength dipoles.

separation (wavelengths)	*gain (dBi)*	*stacking gain (dB)*
0.000	2.15	0.00
0.125	2.43	0.27
0.250	3.24	1.09
0.375	4.51	2.36
0.500	5.98	3.83
0.625	6.94	4.79
0.750	6.76	4.61
0.875	5.83	3.68
1.000	4.93	2.78
1.250	4.37	2.22
1.500	5.28	3.12
1.750	5.84	3.69
2.000	5.10	2.95

Take another look at the two half-wavelength dipoles of Fig. 4.1. The two (broadside) dipoles separated by one-quarter wavelength will produce a gain increase over one dipole. This separation or stacking gain can be easily calculated by using known self- and mutual impedances; the result is shown in Table 4.1, and also in Fig. 4.3 where the overall gain is plotted against the separation. Note that the gain improvement peaks when the separation is about 5/8 wavelengths, leading to the often-quoted (but generally incorrect for other than single elements) "optimum stacking separation."

Fig. 4.3 can be understood qualitatively by remembering that the effective aperture cross section area for a single half-wavelength dipole is about 0.13 square wavelengths. Thus, when the separation between the dipoles is large

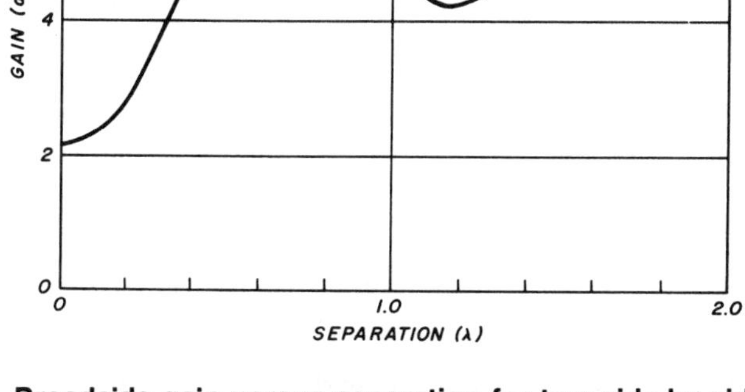

Fig. 4.3 — Broadside gain versus separation for two side-by-side dipoles.

enough, the aperture areas are basically independent (and thus additive), leading to an effective gain improvement of a factor of 2 (3 dB). For smaller separations, however, the aperture areas overlap. In this overlap region both constructive and destructive interference can occur; for very small separations the combined aperture areas reduce to that of a single dipole, but for some intermediate regions interference occurs making the gain improvement more than a factor of 2. This constructive interference is due to a favorable reduction in the driving-point resistance of the dipole, which is a direct result of the behavior of the real part of the mutual impedance of the two dipoles.

Another qualitative way of understanding this entire phenomenon is to view the (transmitting) dipoles as an excitation of a single vertical aperture. Broadening this aperture by separating the dipoles is tantamount to narrowing the *H*-plane pattern, which will increase the gain. When the dipole separation becomes large enough, however, the quasi-uniform illumination disappears and the vertical aperture acts like two dependent (small) apertures giving rise to a diffraction pattern in the *H*-plane with maxima corresponding to a gain improvement of exactly a factor of 2. In any case, the actual calculated values of gain are shown in Table 4.1. Note that for the quarter-wavelength separation of Fig. 4.1, the gain of the stacked dipoles is 3.24 dBi, or just 1.09 dB above that for a single half-wavelength dipole. This gain increase is all due to beam pattern narrowing in the *H*-plane; the *E*-plane pattern beamwidth remains the same as that for a single half-wavelength dipole.

The square loop of Fig. 4.2 is very similar to the stacked dipoles of Fig. 4.1, but there are two significant changes. Because of the truncated or shortened elements, the driving-point impedances of the elements are reduced and the *E*-plane pattern width is somewhat increased, resulting in a somewhat reduced gain. To calculate these two results I will use the same method outlined by Kraus (pages 139-143).[1] Kraus's calculation applies to a thin radiating element which is not capacitance loaded; to make the calculation apply to a single truncated element fed at the maximum current points, the limits of

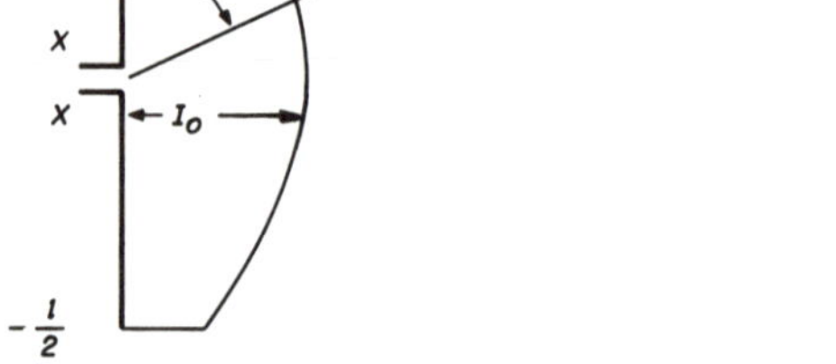

Fig. 4.4 — Current relationships for the symmetrical, thin center-fed truncated antenna of length *l*. This is the relationship necessary to determine driving-point impedance and *H*- and *E*-plane patterns of the square quad loop.

[1]This and other footnotes appear at the end of this chapter.

integration must be suitably altered. Fig. 4.4 shows the essential geometry.

The retarded value of the current at point z on the antenna referred to a point at distance, S, is

$$I = I_0\,[\cos(2\pi z/\lambda)]e^{j\omega(t - S/c)} \tag{4.1}$$

where the quantity in brackets is the form factor for the current on the antenna. Following the Kraus development, the antenna can be viewed as a string of infinitesimal dipoles of length dz. The far fields, dE_θ and dH_ϕ, at a distance, S, from an infinitesimal dipole, dz, are

$$dH_\phi = j\,\frac{I\sin\theta}{2S\lambda}\,dz \tag{4.2}$$

$$dE_\theta = 120\pi dH_\phi \tag{4.3}$$

where ϕ is the angle around the z axis. The total field from the entire antenna is then

$$H_\phi = \int_{-l/2}^{l/2} dH_\phi \tag{4.4}$$

From (4.1), (4.2), and (4.4),

$$H_\phi = j\,\frac{I_0\sin\theta}{2\lambda}\,e^{j\omega t}\int_{-l/2}^{l/2}\frac{1}{S}\cos(2\pi z/\lambda)\,e^{-j\omega S/c}dz \tag{4.5}$$

Note that

$$S = r - z\cos\theta \tag{4.6}$$

and also note that at long distances the amplitude of S is the same as the amplitude of r, so that we may write:

$$\mathrm{H}_\phi = j\,\frac{I_0\sin\theta}{2\lambda r}\,e^{j\omega(t - r/c)}\int_{-l/2}^{l/2}\cos(2\pi z/\lambda)e^{j(\omega z/c)\cos\theta}\,dz \tag{4.7}$$

Let $\beta = \omega/c = 2\pi/\lambda$; then (4.7) may be rewritten:

$$H_\phi = j\,\frac{\beta I_0\sin\theta}{4\pi r}\,e^{j\omega(t - r/c)}\int_{-l/2}^{l/2} e^{j\beta z\cos\theta}\cos(\beta z)\,dz \tag{4.8}$$

Since

$$\int e^{ax}\cos bx = \frac{e^{ax}[a\cos bx + b\sin bx]}{a^2 + b^2}$$

then with $a = j\beta\cos\theta$ and $b = \beta$, (4.8) becomes:

$$H_\phi = j\,\frac{\beta I_0\sin\theta}{4\pi r}\;e^{j\omega(t-r/c)}\;e^{j\beta z\cos\theta}\,\frac{j\beta\cos\theta\cos\beta z + \beta\sin\beta z}{\beta^2\sin^2\theta}\Bigg|_{-l/2}^{+l/2} \tag{4.9}$$

Evaluating this expression at both limits and collecting terms:

$$H_\phi = j\,\frac{[I'']}{2\pi r}\,F(\theta) \tag{4.10}$$

where

$$[I''] = I_0 e^{j\omega(t-r/c)} \tag{4.11}$$

and

$$\mathrm{F}(\theta) = \frac{1}{\sin\theta}[\cos(\beta\frac{l}{2}\cos\theta)\sin(\beta\frac{l}{2}) - \cos\theta\sin(\beta\frac{l}{2}\cos\theta)\cos(\beta\frac{l}{2})] \tag{4.12}$$

$F(\theta)$ is often referred to as the field pattern factor. Thus the far fields of the truncated element can be written:

$$H_\phi = j\,\frac{[I'']}{2\pi r}\,F(\theta) \tag{4.13}$$

and

$$E_\theta = j\,\frac{60[I'']}{r}\,F(\theta) \tag{4.14}$$

Note that if there is no truncation ($l = \lambda/2$) (4.12) reduces to the well-known expression for a half-wavelength dipole:

$$F(\theta) = \frac{\cos(\frac{\pi}{2}\cos\theta)}{\sin\theta}$$

Kraus has shown that the self-resistance, R_{11}, of such a linear element can be computed by equating the integral of the Poynting vector over a large sphere (total power radiated) to the driving-point (current maximum) total power supplied. The result is (see Kraus, ref. 1, equation 5-90, page 143):

$$R_{11} = 60\int_0^\pi F^2(\theta)\sin\theta\, d\theta \tag{4.15}$$

For the truncated element it is now a simple matter to insert the value of $F(\theta)$ from (4.12) into (4.15) and integrate. The integration is quite easily done numerically with a simple computer program. The result of such an integra-

Table 4.2 — Self-resistance and gain versus length for a truncated dipole element.

length (λ)	*resistance* (ohms)	*directivity*	*gain* (dBi)
0.05	2.0	1.502	1.77
0.10	7.6	1.510	1.79
0.15	16.3	1.522	1.82
0.20	27.0	1.537	1.87
0.25	38.5	1.557	1.92
0.30	49.8	1.578	1.98
0.35	59.6	1.599	2.04
0.40	67.0	1.619	2.09
0.45	71.6	1.635	2.13
0.50	73.1	1.641	2.15

tion is shown in Table 4.2 where the self-resistance of a truncated dipole element of length l (in terms of wavelengths) is shown.

The directivity or gain can also be easily computed. Once $F(\theta)$ is given, the directivity is simply the ratio of the maximum value of $F^2(\theta)$ to the average value over the entire 4π solid angle. That is, the directivity D is:

$$D = \frac{4\pi F^2(\theta = \pi/2)}{2\pi \int_0^{\pi} F^2(\theta)\sin\theta d\theta} \tag{4.16}$$

or from (4.15)

$$D = \frac{120}{R_{11}} F^2(\theta = \pi/2) = \frac{120}{R_{11}} \sin^2(\beta \frac{l}{2}) \tag{4.17}$$

The calculated directivity, D, and the corresponding gain, expressed in dBi, are also listed in Table 4.2.

Note that for small truncations where l is only slightly shorter than one-half wavelength, there is not much reduction in self-resistance or in directivity; this is to be expected because the small ends of the dipole which are truncated carry little current, so do not contribute greatly to element performance. For heavy truncation, however, both self-resistance and directivity decrease significantly; the limiting case where the length goes to zero is the well-known infinitesimal dipole whose directivity is just 1.500 (1.761 dBi).

To compute the driving-point resistance of the full square loop it is necessary to know the mutual resistance between two (truncated) elements separated by one-quarter wavelength. From quite fundamental considerations Hurwitz[2] has shown a mathematical expression for the real part of the complex mutual impedance between two elements, R_{21}:

$$R_{21} = 60 \int_0^{\pi} F^2(\theta)\sin\theta J_0(\beta S\sin\theta)d\theta \tag{4.18}$$

where $F(\theta)$ is the pattern function and J_0 is the Bessel function whose argument is the product of element separation S and $\beta\sin\theta$. If S is measured in units of wavelength the argument is simply $2\pi S\sin\theta$. Note that for very small separations J_0 approaches unity; thus, for very close separations $R_{21} \approx R_{11}$.

Shown in Table 4.3 are values of mutual resistance versus separation for truncated elements. Note that the mutual resistance behaves very similarly to the values for full half-wavelength dipoles (see Chapter 1) but the magnitudes are much smaller; a careful comparison shows that the reduction factor varies somewhat with separation.

The properties of the square loop can now be computed. Table 4.2 shows that the gain of a *single* (truncated) quarter-wavelength element is 1.92 dBi and that $R_{11} = 38.5$ ohms. For the full loop of Fig. 4.2 there are two "elements;" if they are both equally driven, they will produce the same far-field strength and power density as a single driven "element" carrying twice the current. However, the driving-point resistance of each one of the driven elements, when both elements are equally excited, is $R_{11} + R_{21}$. Therefore, the total input power (both elements) is:

Table 4.3 — Mutual resistance versus separation for one-quarter wavelength truncated elements.

separation (λ)	*resistance* (ohms)	*separation* (λ)	*resistance* (ohms)
0.00	38.5		
0.05	37.8	0.80	−9.8
0.10	35.5	0.85	−7.2
0.15	32.0	0.90	−4.2
0.20	27.3	0.95	−1.1
0.25	21.7	1.00	1.7
0.30	15.7	1.10	6.1
0.35	9.6	1.20	7.7
0.40	3.7	1.30	6.5
0.45	−1.7	1.40	3.2
0.50	−6.2	1.50	−0.8
0.55	−9.6	1.60	−4.0
0.60	−11.9	1.70	−5.5
0.65	−13.0	1.80	−4.8
0.70	−12.9	1.90	−2.5
0.75	−11.8	2.00	0.4

$$W_2 = 2I^2(R_{11} + R_{21}) \qquad 4.19)$$

For the single element alone (same far field) the total power is:

$$W_1 = (2I)^2 R_{11} \qquad (4.20)$$

If the directivity of a single element is designated as D_1 and the directivity of the full loop as D_2, then

$$D_2 = D_1 W_1/W_2 = 2D_1/(1 + R_{21}/R_{11}) \qquad (4.21)$$

From Tables 4.2 and 4.3 values for the square loop are: $R_{11} = 38.5$ ohms; $R_{21} = 21.7$ ohms; D_2/D_1 becomes 1.279 (1.07 dB). Note that this stacking gain of the two truncated elements is nearly identical to the stacking gain of half-wavelength dipoles at a separation of one-quarter wavelength (1.09 dB, see Table 4.1); the difference is due only to the details of mutual resistance. The driving-point resistance, R, of the total loop, of course, is twice the value for a single element (when both are driven):

$$R = 2(R_{11} + R_{21}) = 120.4 \text{ ohms} \qquad (4.22)$$

Since D_1 (see Table 4.2) is 1.557, $D_2 = 1.991$ (2.99 dBi).

These properties of the square loop have been obtained rather rigorously; the main assumption in the model is total neglect of far-field radiation from the vertical sections, which are assumed to act as capacitance sinks for the current at the ends of the horizontal radiating segments. Moreover, the assumption is made that the current distribution on the horizontal segments is strictly sinusoidal; this is valid for very thin elements. Most loops are built with wire, which is thin compared to a wavelength, so one can be quite confident of the model.

It is now easy to understand qualitatively the radiation pattern of the loop. The *H*-plane profile does narrow, compared with a single linear element, because of the "illumination" of a wider vertical aperture; quantitatively, this profile narrowing is almost the same for truncated loop "elements" and the equivalently separated half-wavelength dipole. The *E*-plane profile, however, is not as narrow for the truncated element as for a half-wavelength dipole; this factor accounts for the somewhat reduced directivity (see Table 4.2). The loss in gain is about 0.23 dB for the quarter-wavelength truncated element.

Other Driven Loops

Before I consider a multiloop system I will briefly discuss other forms of driven loops. First, I will determine the performance of the square loop when it is driven at a point other than the center of the bottom segment. Fig. 4.5 shows the square loop with three different feed points: the center of the bottom leg *xx*, the center of the right vertical leg *yy*, and at an arbitrary point, *zz*, placed at a counter-clockwise distance *d* (from *xx*) around the loop. It has already been shown that if excitation is applied at *xx* (with *yy* and *zz* shorted), the loop will produce an horizontally polarized far field; its gain is 2.99 dBi. Similarly, if excitation is applied at *yy* (with *xx* and *zz* shorted), only vertically polarized far-field radiation will occur; gain is again just 2.99 dBi. Note that excitation *xx* and *yy* are basically independent, i.e., unit current flow due to excitation at *xx* has a null in current at *yy* and vice versa.

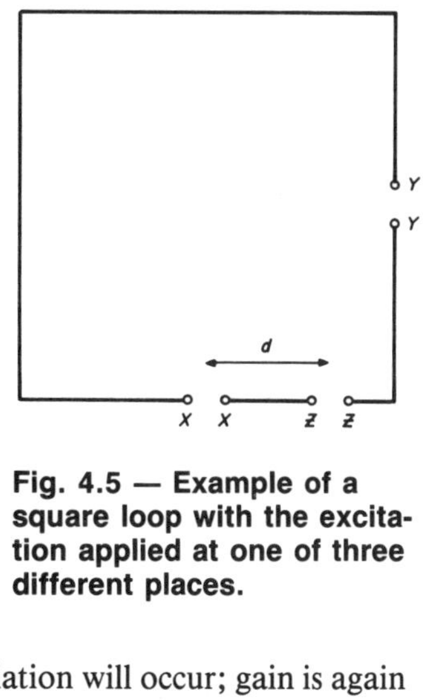

Fig. 4.5 — Example of a square loop with the excitation applied at one of three different places.

If we now excite the loop at *zz* with the same unit current, $I_0\cos\omega t$, it is easy to see that current flow at *xx* is just $I_0\cos(\beta d)\cos\omega t$ and at *yy* is $I_0\sin(\beta d)\cos\omega t$. These currents will produce orthogonal far fields which must be added vectorially to obtain the total far field; the total far field has exactly the same magnitude as that which is produced by the same unit current at *xx* alone. In other words, the square loop gain and the drive-point resistance are totally independent of the feed point; only the polarization changes (from totally horizontal if the feed is at *xx* to totally vertical if the feed is at *yy*). This simple theorem can be easily proved by the same type of argument for any equilateral one-wavelength loop.

It is now interesting to consider moving the feed point *zz* to the lower right-hand corner of the square loop. Since we are considering free space loop gain for which rotation is unimportant, it is clear that this configuration produces *exactly* the same result as the familiar square diamond loop fed at the bottom or top corner. Thus we now know that the quad and the diamond loop have *exactly* the same gain and *exactly* the same drive-point resistance. Similarly, the gain and drive-point resistance of *any* equilateral (one-wavelength) loop is *totally* independent of the position of the drive point on the loop.

I shall now return to the horizontally polarized square of Fig. 4.2. We have shown that a square loop (relative to a halfwave dipole) has a somewhat enhanced gain (+0.84 dB), made up of an increase (+1.07 dB) due to the vertical separation of the radiating segments (*H*-plane narrowed somewhat) and a decrease (−0.23 dB) due to the truncated radiating segments (*E*-plane broadening). Let's now explore the performance of a rectangular one-wavelength loop; Fig. 4.6 shows some examples.

Rectangle A is a wide but low loop which is recognized as a folded dipole loop; B is a narrower and higher loop than the square, and C is a high but very narrow loop. For each of these loops the sum of the width and height is constrained to be 0.5 wavelengths. We are now in a position to compute

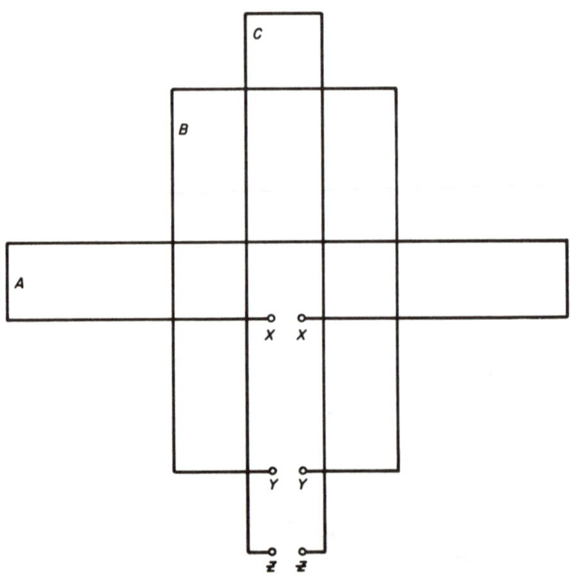

Fig. 4.6 — Diagram of three different rectangular configurations. In each case the feed point is located for horizontally polarized radiation.

Table 4.4 — Directivities, gain and self-resistance for one wavelength rectangular loops with various width-to-height ratios.

width (wavelengths)	*height (wavelengths)*	D_1	D_2/D_1	D_2	*gain (dBi)*	*resistance (ohms)*
0.05	0.45	1.502	2.073	3.114	4.93	3.8
0.10	0.40	1.510	1.815	2.741	4.38	16.7
0.15	0.35	1.522	1.596	2.430	3.86	40.8
0.20	0.30	1.537	1.419	2.181	3.39	76.0
0.25	0.25	1.557	1.279	1.991	2.99	120.5
0.30	0.20	1.578	1.172	1.850	2.67	169.8
0.35	0.15	1.599	1.094	1.750	2.43	217.7
0.40	0.10	1.619	1.041	1.686	2.27	257.5
0.45	0.05	1.635	1.010	1.652	2.18	283.6
0.50	0.00	1.641	1.000	1.641	2.15	292.5

loop performance. Table 4.4 shows the results of calculations for rectangular loops by the same methods as used for the square. D_2/D_1 is the H-plane directivity increase due to the vertical separation of radiating segments (see (4.21)), while D_1 is the directivity of a single horizontal radiating segment (see Table 4.1). Note that the limiting case of the folded dipole (width = 0.5, height = 0) shows a gain of 2.15 dBi (identical to a single halfwave dipole) and a driving-point resistance of 292.5 ohms (just four times that of a single halfwave dipole). Since reactance effects of the full loop are double that of a single "element," the Q of the folded dipole will be just one-half the Q of a single halfwave dipole, leading automatically to a bandwidth twice as large. As one reduces the loop width from this limiting case the gain increases monotonically. However, this favorable increase in gain is automatically accompanied by a significant reduction in driving-point resistance; since reactance effects are essentially identical for all loops, the circuit Q increases commensurately. Thus the potential gain obtainable from high narrow loops is always offset by unfavorably high values of Q and corresponding narrow bandwidths.

Let us now consider other loop shapes. For equilateral shapes, we have already seen that gain and driving-point resistance are independent of feed point position on the loop. Moreover, a reasonably rigorous solution has been obtained for a square loop. Because of the independence of properties on feed point position, which is equivalent to independence of loop rotation with feed point fixed at the bottom, it seems reasonable to assume that gain from all equilateral loops is approximately equal to that of a circular loop having the same enclosed area. This in turn can be equated to an equivalent square of the same area. Such an intuitive model, together with a model of how E-plane directivity varies with element length (see Table 4.2) and how H-plane directivity varies with element separation (see Table 4.1), allows a reasonable estimate of equilateral loop gain. Table 4.5 lists these values for several equilateral loops. The values for the square are the ones already com-

Table 4.5 — Estimated directivities, gain and self-resistance for one wavelength equilateral loops.

equilateral loop type	*sides*	*equivalent square loop side length (wavelengths)*	D_1	D_2/D_1	D_2	*gain (dBi)*	*resistance (ohms)*
triangle	3	0.219	1.545	1.205	1.862	2.70	104.7
square	4	0.250	1.557	1.279	1.991	2.99	120.5
pentagon	5	0.262	1.562	1.309	2.044	3.10	126.4
hexagon	6	0.269	1.565	1.324	2.071	3.16	129.3
octagon	8	0.275	1.567	1.338	2.097	3.22	132.0
circle	∞	0.282	1.570	1.356	2.129	3.28	135.3

puted (see Table 4.4); all others are estimated by this simple model. It is quite easy to see that the popular triangle or delta loop is slightly lower in gain (by 0.3 dB) than the square (quad or diamond) and that the loop with the highest gain is a circle. However, note that the gain of the circular loop is estimated as 3.28 dBi which is 1.13 dB larger than that of a halfwave dipole. This is not quite as large an increase as the approximately 2 dB which was quoted by Lindsay, but I believe this discrepancy is probably not outside of the experimental accuracy range of Lindsay's measurements combined with the estimation accuracy range of the simple model I have used.[3]

Multiloop Arrays

I shall now consider an array of square loops with both driven and parasitic loop elements. From what we have just seen, the array should behave just about like a two-array stack of equivalent Yagi antennas, separated (vertically) by one-quarter wavelength and having elements bent together to form the individual square loops. However, to proceed with a computation of the properties of the parasitic loop array, we must know all complex self- and mutual impedances of the truncated elements. Up to this point it has been possible to rather rigorously describe the properties of a driven loop because only the real part of self- and mutual impedances were required to obtain gain, driving-point resistance, and pattern information (both in the *E*- and *H*-planes).

When parasitic elements are involved, the imaginary parts of the impedance terms are required. Computation of these reactances is a non-trivial exercise, and I am unaware of any rigorous procedure for carrying out such a calculation for truncated elements. As far as self-impedance is concerned, the reactive value is controlled entirely by the "tuning" of the loop; that is, the relationship of wavelength to loop circumference and the effective loop Q. The complex impedance has been calculated for a linear nearly halfwave thin cylindrical element.[4] The method involves treating the metallic cylinder as a boundary-value problem (tangential components of electric field are made to vanish at every point on the conductor surface), from which an integral equation is derived. Approximate solutions of this integral equation yield the current distribution on the cylindrical element from which the input impedance, including the imaginary component, was derived.

I have been unable to find an equivalent rigorous boundary-value calculation for the square loop; thus we do not yet have the basis for calculating the precise reactance for a nearly one-wavelength square loop. However, it is possible to at least estimate the loop Q (but not its precise resonant frequency) by remembering that reactance changes with frequency are due to the effective inductance and capacitance of the antenna, that is, near-field stored energies, whereas the resistance (radiation) has to do with the far field. Truncating the halfwave dipole changes the geometry of the element but hardly affects its (central) inductance or (end) capacitance. It would be reasonable, therefore, to assume the square loop contains essentially the same total reactive impedance changes as two halfwave dipoles. Thus, following the argument made in Chapter 1, we may write for the loop self-impedance:

$$Z_{11} = R_{11} + X_{11} \approx R_{11}[1 + j2Q(F - F_R)] \tag{4.23}$$

But (empirically)

$$R_{11}Q \approx 430.8 \log K - 339 \tag{4.24}$$

so that

$$Z_{11} \approx 120.5 + j(861.6 \log K - 678)(F - F_R) \tag{4.25}$$

As in Chapter 1, F is the normalized frequency and F_R is the normalized resonant frequency of the loop. Loop Q is readily estimated from (4.24) and (4.22). The only remaining problem is determination of F_R. Although there is no rigorous way of calculating F_R from basic principles, it is significant to note that the region where the ends of the two dipoles are joined must have electric (capacitive) fields at right angles to the conducting cylindrical element, exactly like those of an infinitely long straight cylinder near a voltage node. This observation implies that there should be a negligible "end effect" at the capacitive voltage loop, and, therefore, F_R should be very close to the frequency at which the total node circumference is just one wavelength.

When we consider the imaginary part of the mutual impedance between loop halves or "elements," another computational complication arises. At long distances or separations, the imaginary part of the mutual reactance, X_{21}, must be (except for a phase shift) simply related to the real part, R_{21}, and this relationship should be unaffected by the precise "tuning" of the "elements." However, at very small separation X_{21} must approximate the value of X_{11}, which is fundamentally fixed by circuit Q and resonant frequency and not at all by R_{11}. How to correctly represent X_{21} at all intermediate spacings has not, to my knowledge, been solved quantitatively. For this reason I will model the loop array as the equivalent Yagi stacked

Table 4.6 — Element lengths and positions for representative beams. Radius of all elements = 0.00052599 wavelengths.

number of elements	element lengths (wavelengths) refl.	driver	dir.	boom length (wavelengths)	element spacing (wavelengths)	Fig. 4.7 curve
2	0.4937	0.4705	—	0.15	0.150	2
3	0.4980	0.4896	0.4690	0.25	0.125	3
6	0.4953	0.4803	0.4481	0.75	0.150	6
7	0.4936	0.4762	0.4466	1.25	0.208	7

at a spacing of one-quarter wavelength, then apply necessary corrections to directivity and gain. Compared with stacked (halfwave element) Yagis, the loop array will have 0.23 dB less directivity and gain (all due to *E*-plane pattern broadening).

It is interesting to calculate the free space "stacking gain" of Yagis as a function of vertical separation. I chose as representative boomlengths and number of elements the following simplistic Yagis, which were found to have excellent free space properties (see Chapter 2): a two-element beam on a 0.15 wavelength boom, a three-element beam on a 0.25 wavelength boom, a six-element beam on a 0.75 wavelength boom, and a seven-element beam on a 1.25 wavelength boom. Table 4.6 lists characteristics of these beams.

Fig. 4.7 shows the computed gains of these stacked configurations as a

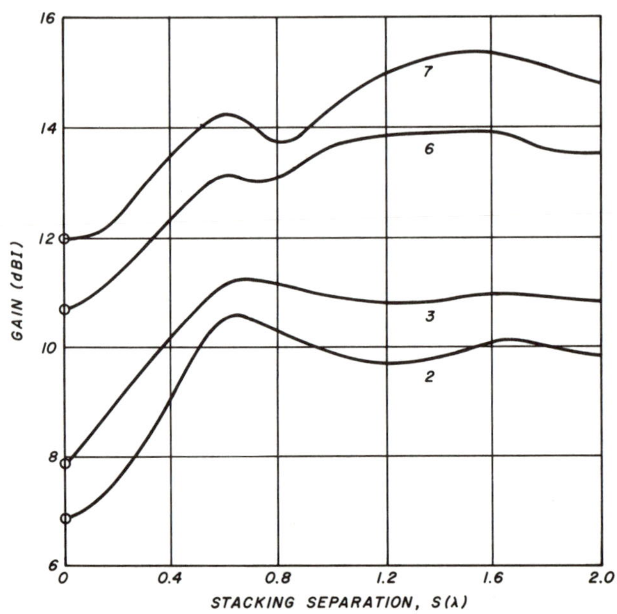

Fig. 4.7 — Stacking gain versus separation for representative two, three, six, and seven-element Yagis.

function of the stacking separation. Note that the variation in gain with separation is somewhat different for each case and also somewhat different that the case for dipoles shown in Fig. 4.3. Two things seem to be occurring as the separation increases. For small separations, the capture area of one Yagi essentially overlaps that of the other Yagi; therefore, the total capture area for both is essentially the same as for one alone. As the separation increases, the total capture area increases and ultimately doubles if the spacing is large enough. For the larger Yagis (where the original capture area for one Yagi is large), it is easy to see that to realize a given separation gain, the spacing must be relatively larger than for a small Yagi or especially a dipole. In other words, the transition from the gain of one Yagi to the doubling of gain (3 dB increase) for two Yagis requires *larger* separation for *larger* Yagis. In addition to this rather gradual gain increase due to separation of capture areas, Fig. 4.7 suggests that the constructive interference due to mutual impedances noticed for the dipoles in Fig. 4.3 also persists in stacked Yagis. An increase in gain is noticeable at separations of 0.6 and 1.6 wavelengths. Qualitatively, Fig. 4.7 shows a combination of these two effects; first, the constructive impedance gain increase at particular separations and second, the capture area separation gain increase which takes place more slowly with large Yagis.

It is interesting to compare the computed gain increase of the seven-element beam with the experimental values reported by Viezbicke in his Fig. 11A.[5] Table 4.7 shows a comparison of my computed results with Viezbicke's published experimental values. The comparison is not totally valid because Viezbicke's seven-element beam is not the same beam I have used in Fig. 4.7. The differences cannot be quantified, since Viezbicke failed to include specifications for his seven-element beam. Nevertheless, Table 4.7 shows good qualitative agreement and even fair quantitative agreement (close to Viezbicke's stated experimental accuracy of 0.5 dB).

Table 4.7 — Measured and computed stacking gain versus spacing for seven-element Yagis.

spacing (wavelengths)	*Viezbicke (dBi)*	*computed (dBi)*
0.38	0.80	1.3
0.57	1.58	2.15
0.78	1.36	1.8
0.99	1.90	2.35
1.20	2.34	3.0
1.40	2.53	3.25
1.61	1.93	3.3

Computations represented by the data shown in Fig. 4.7 cannot really be carried down to very small separations with any confidence because the mathematical model uses mutual impedances of full halfwave elements. When reactive parasites get very close together their mutual impedance has an imaginary component quite different from that of halfwave dipoles.

I shall now examine the stacking gain increases for a separation of one-quarter wavelength. Table 4.8 shows these increases for the four computed cases (dipole data from Fig. 4.3). Note that the quarter-wavelength stacking

Table 4.8 — Stacking gain for two Yagis one-quarter wavelength apart, and estimated gain of a quad over a single Yagi, for various number of elements.

number of elements	*boom length (λ)*	*Yagi stacking gain (dB)*	*quad gain over single Yagi (dB)*
1	0.00	1.09	0.86
2	0.15	1.03	0.80
3	0.25	1.38	1.15
6	0.75	0.84	0.61
7	1.25	0.65	0.42

gain for very large Yagis is *not* as large as that for dipoles; this can easily be understood because of the (relatively) smaller increase in capture area. One might expect this gain increase to fall monotonically from the value of 1.09 dB (for dipoles) as the array length is increased, however, Table 4.8 shows the actual increase to vary somewhat due to the detailed way in which impedances vary.

We are now in a position to estimate the performance of a loop array. If instead of the dipole elements of the stacked Yagi arrays I use (bent) elements connected in square loops, I will make a loop array in which all conditions remain about the same *except* that the *E*-plane pattern is broadened and the gain is correspondingly reduced due to the truncated "elements." The gain performance of the loop array is therefore about 0.23 dB lower than Yagis stacked one-quarter wavelength apart. These estimations are also shown in Table 4.8.

It must be evident by now that an array of square loops is very much like an equivalent Yagi. Its overall gain is expected to be somewhat higher than the Yagi because of individual loop gain, but not by very much. The *E*-plane pattern is slightly broader due to the current distribution on the truncated "elements," whereas the *H*-plane pattern is slightly narrower due to the "stacking gain" of the loops. Because of this similarity of Yagi and loop arrays, it is plausible that one can intermix loops and linear elements to provide a hybrid structure of roughly equal performance. Such a hybrid is known as a "quagi"; if properly constructed, it should provide a pattern and gain intermediate between a similar quad and a similar Yagi. There are obviously an enormous variety of possible quagi configurations; there will remain for some time a challenge to the quagi designer to determine preferred configurations and best dimensions for all radiating elements.

Summary

A number of interesting conclusions have been reached regarding antenna arrays constructed with conducting loops roughly one wavelength in circumference.

1) A single (driven) loop will provide a free space gain somewhat larger than that of a halfwave dipole. The gain increase comes about through a narrower *H*-plane pattern and a slightly broader *E*-plane pattern.

2) Loop gain varies significantly with shape. For rectangular loops fed in the center of the lower horizontal segment, gain depends on the ratio

of width to height, varying from 2.15 dBi (equal to a halfwave dipole) to about 5 dBi (3.8 dB above a halfwave dipole) as the ratio approaches zero.

3) For equilateral loops, gain depends only on the number of sides (see Table 4.5). For the square loop, the free space gain calculated rather rigorously is 2.99 dBi and driving-point resistance is 120 ohms. It is significant that these properties of the square loop are totally independent of the feed point on the square. As an example, the gain and drive-point resistance for a square loop is exactly the same as that of the diamond square.

4) The free space gain of a loop array is estimated to be somewhat higher than that of an equivalent (single) Yagi. Calculations show this difference depends somewhat on the particular array (see Table 4.8), but ranges from about 1 dB for short arrays to less than 0.5 dB for long arrays.

5) Quagi configurations are expected to show performance figures between those of an equivalent Yagi and equivalent quad.

6) A rigorous theory does not yet exist for self- and mutual quad loop reactances. Consequently, quad (and quagi) parasitic loops must be experimentally adjusted for correct resonant frequency. It is unlikely that such experimental adjustments can be made with the same precision and confidence that linear element lengths can now be specified by present theory. Therefore, the slight gain advantage of the quad over the Yagi shown in Table 4.8 may well disappear in practice.

NOTES:

[1]Kraus, John D., *Antennas*, New York: McGraw-Hill, 1950.
[2]Hurwitz, H., W2HH, Private communication.
[3]Lindsay, J., "Quads and Yagis," *QST*, May, 1968.
[4]Hallen, E., "Theoretical Investigations into the Transmitting and Receiving Qualities of Antennae," *Nova Acta Uppsala (Sweden)*, Series IV, Volume 11, Number 4, pp 3-44, 1938.
[5]Viezbicke, P. "Yagi Antenna Design," *NBS Technical Note 688*, U.S. Department of Commerce, Washington, D.C., December, 1976.

CHAPTER 5

THE EFFECTS OF GROUND

So far the antennas I've discussed have been considered in free space conditions. In the real world, however, the performance of an antenna is profoundly affected by the proximity of the ground. The ground can be beneficial in some circumstances and detrimental in others; it is important to understand ground effects and how to use the presence of ground advantageously.

Unfortunately, the ground near an antenna is not easily characterized; ordinarily, it is complex in shape with highly variable radio frequency properties. The ideal "ground" will consist of a flat plane of material with high electrical conductivity, usually approximated quite well by the surface of the ocean. Actual ground, however, has a much lower conductivity and lower dielectric constant than salt water and is ordinarily far from flat. Values of electrical conductivity range from about 4 mhos/meter for salt water to as low as 10^{-3} mhos/meter for typical residential areas; corresponding values of dielectric constant range from about 80 for water to 5 for typical residential areas.

The presence of ground alters antenna performance in two ways. First, it changes the normal free space pattern (in the illuminated hemisphere) by adding (vectorially) a reflected pattern. The combination of direct and reflected radiation fields produces regions of enhanced gain, but at the expense of reduced gain in other regions. Second, it alters the antenna itself; currents that flow in the ground surface couple back into the antenna and change all element currents. In other words, an antenna becomes a somewhat different antenna when placed near the ground.

Reflections From A Plane Ground

Let us first consider the reflection from a plane ground surface: the reflected wave can be characterized by a reflected amplitude and a reflected phase shift, both relative to the original wave. For horizontally polarized radiation the reflected amplitude remains close to unity and the phase shift is close to 180 degrees for all incoming (and therefore reflected) angles with respect to the surface. The phase shift is due to the fact that if the conductivity of the surface is high, Maxwell's equations require the tangential E-field at the surface to vanish. The limiting values only occur if the conductivity of the plane surface is infinite; however, for all practical values of ground conductivity, the reflected amplitude and phase shift remain reasonably close to the limiting values.

By contrast, the reflection of vertically polarized radiation is much more complicated. In this case, the limiting value of the relative reflected wave amplitude (for infinite conductivity where the electric field normal to the surface must be continuous) is also unity, but the phase shift is zero. These values change drastically where the ground has a finite conductivity and dielectric constant and also vary radically with the elevation angle relative to the surface. As the elevation angle is changed from zero to 90 degrees, the reflected amplitude varies from unity down through a minimum (at an angle called Brewster's angle) and slowly back up to a value usually significantly less than unity. At the same time, the phase shift varies from 180 degrees monotonically down to zero. Brewster's angle is a function of the dielectric constant of the surface; for salt water it is about 6 degrees, while for poor ground it rises to about 24 degrees.

Thus, it is easy to see that if one models flat real ground as an infinitely conducting plane, the result should be generally trusted for horizontal polarization, but not for vertical polarization. It is significant that far-field radiation for vertical polarization over real (finite conductivity) ground vanishes at grazing angles. Only when conductivity is truly infinite does grazing angle reflection add to direct antenna radiation.

Ground Curvature Effects

Let me return briefly to the observation that the ground surface near an antenna is rarely flat. Even if this real ground surface were perfectly reflecting (infinite conductivity), the pattern of reflected radiation would be exceedingly complicated. It is useful to imagine a small optical model of the ground surface terrain made with a good shiny reflecting surface and illuminated by a light source (replacing the antenna). From this it is easy to understand that the reflected pattern will show bright spots or glints where converging rays are focused from "dished" (concave) surfaces; in fact, the brightness from such dished areas is potentially much higher than the direct light from the light source itself. If the dished surface is ideal, one can realize enormous brightness gains; the ideal surface is, of course, a parabolic reflector at the correct distance from the source.

Excellent optical searchlights are made with such reflectors. A modern radio frequency example is the large Arecibo dish, which started with a natural, roughly parabolic ground surface and was subsequently improved by high-conductivity, accurately figured reflecting surfaces. To get effective coherent radiation from the entire surface required that the surface maintain an accuracy of a fraction of a wavelength.

The shiny optical model produces a reflected pattern somewhat, but not exactly, like the real ground. The principal difference is due to the very short wavelength of light compared with the size of the surface features of the model. The brightness gain of any well-configured dished area is theoretically proportional to the reflecting (concave) area, but inversely proportional to the square of wavelength. Thus, the optical glints will tend to be small, bright spots; whereas the real radio glints, occurring at the same angles, will be fuzzy, larger, and less bright. Note, however, that the broader radio glint will approximate the average value of the optical glint over the radio glint angle.

Thus, we see that ground around an antenna can not only reflect, but it can focus (and defocus) radiation as well. To get much of a focusing effect, the concave surface must not only be large (compared with a wavelength) and of the correct focal length, but it must also have a surface accuracy of a fraction of a wavelength. This type of surface is not likely to occur naturally at wavelengths in the 10-to-100 meter range, especially if the antenna is situated in a region where the reflecting surface is quite complicated. Nevertheless, for those users fortunate enough to locate their antennas on a relative smooth, concave slope of the right curvature, some significant focusing should take place.

I shall model the ground surface in the conventional way; that is, as a perfectly conducting flat plane. This model should be valid for horizontally polarized radiation at antenna sites where the actual ground is reasonably flat out to distances where specular reflection occurs at the lowest elevation angles of interest. For high angles, the reflection takes place nearly under the antenna, and the ground must be flat in that area to a fraction of a wavelength for the model to apply; this is usually the case.

As the elevation angle is reduced, the reflection point recedes from the antenna location until, at very low angles, it is many wavelengths distant. Under this condition the real ground does not have to be very flat to reflect energy with amplitude and phase coherence; it can, in fact, be quite rough, with variations in height of several wavelengths. This situation is analogous to the well-known optical reflection observed from surfaces that are rough in comparison with a wavelength; one can observe, at grazing angles, nearly specular reflection. A sheet of paper has roughness variations so large compared with the wavelength of light that, at normal incidence, no reflected images can be seen; nevertheless, at grazing angles, one can easily observe specular reflection effects — and even fair images. For these reasons, the model can be expected to be valid for most horizontally polarized antenna systems.

One more point should be mentioned. Because of the finite conductivity of the real ground, the currents that flow are not strictly at the physical surface of the ground but are distributed throughout the top "skin depth" of the ground. This skin depth is usually quite small; as an example, at a frequency of 14 MHz, where the free space wavelength is 21 meters, the skin depth in salt water is less than 0.1 meters; and even for poor ground, where the conductivity is 10^{-3} mhos/meter, the skin depth increases to a value of less than one meter. Therefore, the infinite conductivity plane model of the ground should give quite acceptable results.

Image Models

Besides adding a reflected wave to the space pattern of an antenna, the presence of the highly conducting ground plane changes the properties of the antenna itself. Excitation of the antenna produces a current distribution in the nearby ground plane surface which, in turn, couples mutual voltages back into all antenna elements; these mutual voltages will obviously affect currents in all antenna elements. To model this interaction, it is useful to replace the ground plane conducting surface by an antenna image. This image is located just as far below the original ground plane as the real antenna is located above. For a horizontally polarized antenna, the image is excited equally, but exactly out of phase, with the real antenna.

Because of the geometrical symmetry of the antenna and its image around the original ground plane surface and the opposite excitation, it is easy to see that, at all points on the original ground plane surface, the tangential electric fields vanish (that is, the image field cancels the antenna field). Thus, by means of the image model, we produce exactly the same tangential field at the ground plane coordinates as would be produced by currents flowing in the real conducting ground plane because of antenna excitation alone. The antenna itself cannot distinguish whether a real ground plane conductor or its oppositely excited image exists. Therefore, the ground interaction with the antenna is identical to the image interaction. One can therefore model all antenna properties over the real ground plane by using the antenna and its oppositely excited image in free space.

Note, however, that there is one significant difference between the image model in free space and the real ground plane. In the real ground plane case only a hemisphere is actually irradiated, while in the image model a full sphere is irradiated. Although all fields in the (common) hemisphere will be exactly the same, it is obvious that the total radiated power of the image model will be just twice that of the real ground plane. Therefore, even though antenna element impedances are the same for both situations, gain calculations for the image model must be multiplied by two to obtain gain for the real antenna over the earth.

Propagation Elevation Angles

Before I evaluate detailed antenna properties over earth, I would like to discuss briefly the elevation angles that should be of paramount interest. It is well known that long-distance radio communcations take place primarily by ionospheric F_2 layer reflection (or, more properly, refraction). While the F_2 layer can vary in (virtual) altitude over the earth (250 to 400 km), it is instructive to make a very simple model of this layer as a reflecting shell at an altitude or height of 300 km. A radiated wave at any elevation angle will bounce from this shell and return to earth at the same elevation angle and at a great circle distance which depends only on simple geometry. For a single hop, the maximum range on the earth is limited; thus, communications at very long distances will involve several hops.

Fig. 5.1 plots the elevation angle versus great circle distance for different numbers of hops (up to six). This diagram shows clearly that a given distance can be reached with different discrete elevation angles, or that a given

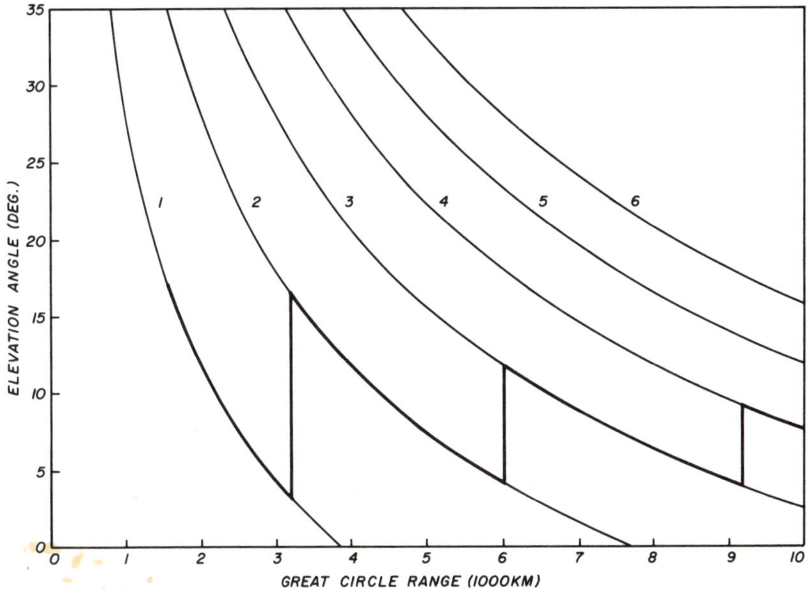

Fig. 5.1 — Diagram of the elevation angle required for propagation over various great circle distances and number hops.

elevation angle arrives back at the earth at discrete distances. To cover all values of distance requires a continuous spread in elevation angle, but the limits of this spread can be narrowed somewhat by taking advantage of different numbers of hops. As an example, all distances beyond 1600 km can be accommodated by the heavy line in Fig. 5.1; that is, elevation angles varying over a range of only 3 to 17 degrees.

It is generally desirable at the higher frequencies to use low angles to minimize attenuation resulting from multiple hops and reflection losses and to ensure ionospheric refraction at the highest frequency. For such frequencies (say 14 to 28 MHz) the range of elevation angles shown in Fig. 5.1 seems quite appropriate. At lower frequencies (7 MHz and below), however, if the propagation path at either end is in daylight (where absorption is high), a higher range of angles (using a greater number of hops) may give a lower overall absorption. The reasoning derived from this simplified model of propagation effects gives expected results not inconsistent with observations reported in *The ARRL Antenna Book* on page 18.[1] However, real propagation is clearly more complicated than is shown in Fig. 5.1.

Kift has shown in an elegant way that long distance propagation involves many propagation modes.[2] He has shown, in measurements made between

[1]This and other footnotes are given at the end of this chapter.

Ascension Island and Slough, England, that measured arrival angles, when complete path ionospheric soundings are known, correlate well with ray-tracing expectations. (Ray tracing can identify actual propagation modes.) His results indicate that elevation angles from 3 to 20 degrees are indeed quite important. He also shows focusing effects of a given mode and the great variety of results that can occur in practice. Thus, in evaluating an antenna system over ground, it is most important to ensure good gain over all lower angles (say 3 to 17 degrees), and for the lower frequencies (7 MHz and below) over even higher angles (up to say 30 degrees). I shall therefore show, in all cases to be presented, a plot of *H*-plane gain as a function of elevation angle.

Antenna Performance Over Ground

To illustrate typical Yagi antenna performances over ideal ground, I shall use as representative Yagi beams a three-element Yagi on a 0.3 wavelength boom and a six-element Yagi on a 0.75 wavelength boom; the basic designs are given in Table 5.1. I have made calculations for each beam at a number of different elevations over ground. The results for the three-element beam

Table 5.1 — Element lengths and positions for representative three and six-element Yagis. All elements are cylindrical with radius = 0.00052599 wavelengths.

	three-element		*six-element*	
element	*length* (λ)	*position* (λ)	*length* (λ)	*position* (λ)
reflector	0.49403	0.000	0.49483	0.000
driver	0.48567	0.150	0.48047	0.150
D1	0.46525	0.300	0.44771	0.300
D2			0.44771	0.450
D3			0.44771	0.600
D4			0.44771	0.750

Table 5.2 — Performance of a representative three-element beam versus height above ground.

					input impedance	
Fig.	*height (wavelengths)*	*gain (dBi)*	*F/B (dB)*	*angle (degrees)*	*resistance (ohms)*	*reactance (ohms)*
5.2A	0.10	9.82	5.6	58	32.6	13.1
5.2B	0.25	11.08	16.8	41	15.0	−4.0
5.2C	0.50	13.19	24.2	27	15.4	1.4
5.2D	0.75	13.84	25.3	19	16.7	−0.9
5.2E	1.00	14.04	25.5	14	15.8	0.7
5.2F	1.25	14.10	32.9	11	16.6	−0.5
5.2G	1.50	14.17	26.5	9	15.9	0.5
5.2H	1.75	14.19	36.1	8	16.5	−0.3
5.2I	2.00	14.24	27.3	7	16.0	0.4
5.2J	2.50	14.24	27.9	6	16.0	0.3
5.2K	3.00	14.25	28.3	5	16.1	0.3
	free space	8.27	31.2	0	16.2	0.1

Table 5.3 — Performance of a representative six-element beam versus height above ground.

Fig.	*height (wavelengths)*	*gain (dBi)*	*F/B (dB)*	*angle (degrees)*	*input impedance resistance (ohms)*	*reactance (ohms)*
5.3A	0.10	11.92	8.7	46	14.6	9.6
5.3B	0.25	13.00	13.2	33	23.2	−6.8
5.3C	0.50	14.85	35.5	23	21.2	−0.6
5.3D	0.75	15.60	21.5	18	20.8	0.7
5.3E	1.00	16.36	37.7	14	21.9	−0.4
5.3F	1.25	16.23	29.3	11	20.5	0.1
5.3G	1.50	16.59	41.7	9	21.8	−0.2
5.3H	1.75	16.44	34.0	8	20.7	−0.1
5.3I	2.00	16.68	45.2	7	21.6	−0.1
5.3J	2.50	16.68	48.3	6	21.6	−0.1
5.3K	3.00	16.70	52.3	5	21.5	−0.1
	free space	10.71	47.0	0	21.2	−0.2

are shown in Table 5.2 and those for the six-element beam in Table 5.3. As a reference, the free space performance for each beam is also listed in these tables. Figures 5.2 and 5.3 show the *H*-plane, or vertical, pattern of each of these cases. It is apparent from these *H*-plane patterns that maximum gain in the forward direction occurs at an elevation angle which is an inverse function of the antenna height; this relationship is tabulated quantitatively in Tables 5.2 and 5.3.

Figures 5.2 and 5.3 also show the antenna gain has a number of lobes; the biggest lobe is the first one (lowest elevation angle). For each succeeding lobe the peak gain is somewhat lower. This reduction in gain is caused by the natural free space directivity of the antenna. The overall pattern is a series of lobes (produced by interference of the direct and reflected waves) essentially modulated by the inherent free space pattern of the antenna. Note that the relative gain reduction at high angles is greater for the (more directive) six-element beam than for the three-element beam. Moreover, a careful analysis of the lobes shows that the maximum point on each lobe is slightly altered by the natural beam directivity. This is shown in Tables 5.2 and 5.3 by the slightly lower elevation angles of the main lobe for the six- versus the three-element beam at a given height above ground.

Thus, we see that the main lobe of an antenna occurs at an angle primarily determined by its height over ground, but secondarily by the natural antenna directivity. This latter effect is most pronounced at low antenna heights; it is also responsible for the relatively poorer gain at these heights. One would ordinarily expect the ground reflection to double the radiated field (or to add 6.02 dB to gain), but if it occurs at a high elevation angle (due to a low antenna height) the original free space antenna gain is significantly lowered (at the same high angle).

The front-to-back ratio is also shown in Tables 5.2 and 5.3. Recall that

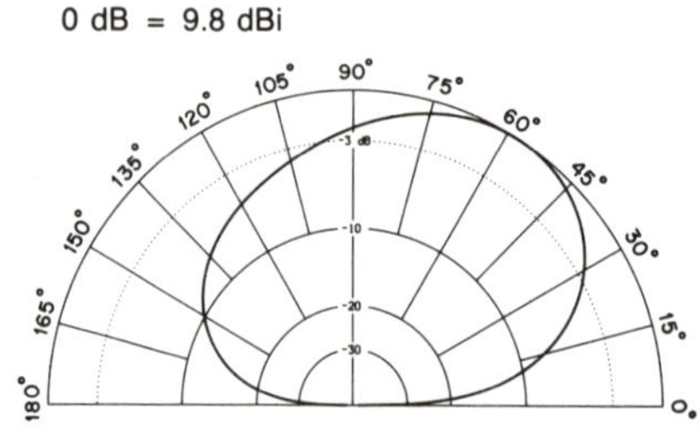

Fig. 5.2A — *H*-plane radiation pattern for a three-element Yagi 0.1 wavelengths above the ground.

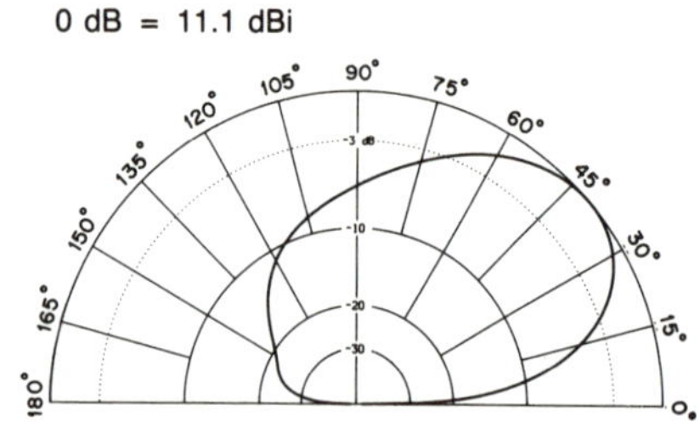

Fig. 5.2B — *H*-plane radiation pattern for a three-element Yagi 0.25 wavelengths above the ground.

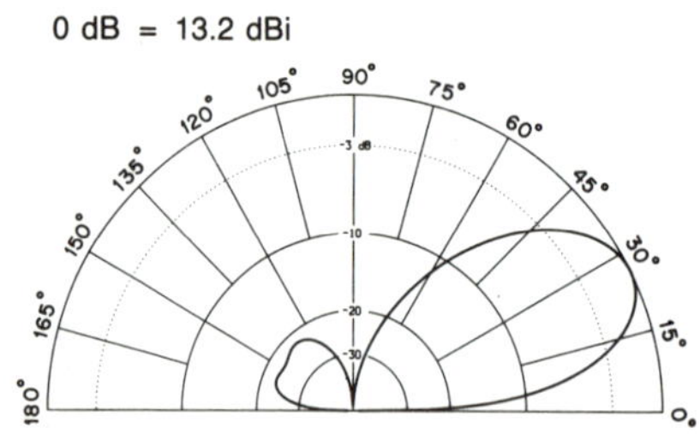

Fig. 5.2C — *H*-plane radiation pattern for a three-element Yagi 0.5 wavelengths above the ground.

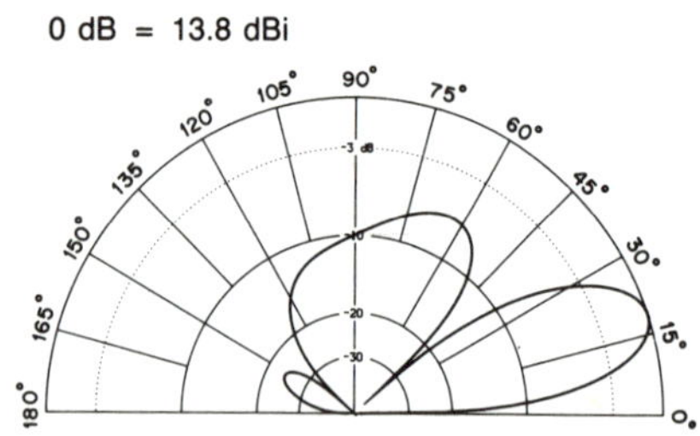

Fig. 5.2D — *H*-plane radiation pattern for a three-element Yagi 0.75 wavelengths above the ground.

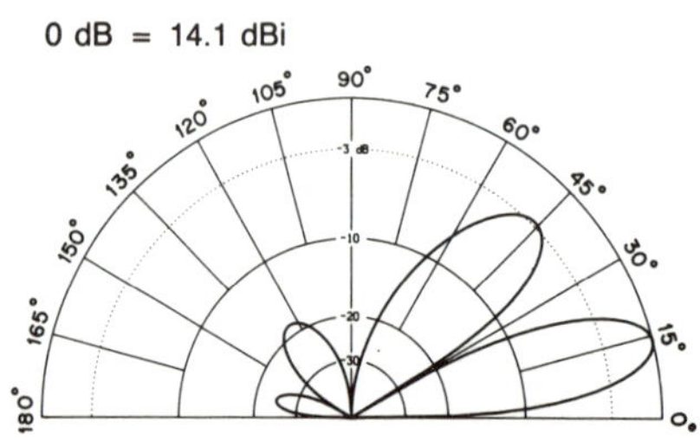

Fig. 5.2E — *H*-plane radiation pattern for a three-element Yagi 1.0 wavelengths above the ground.

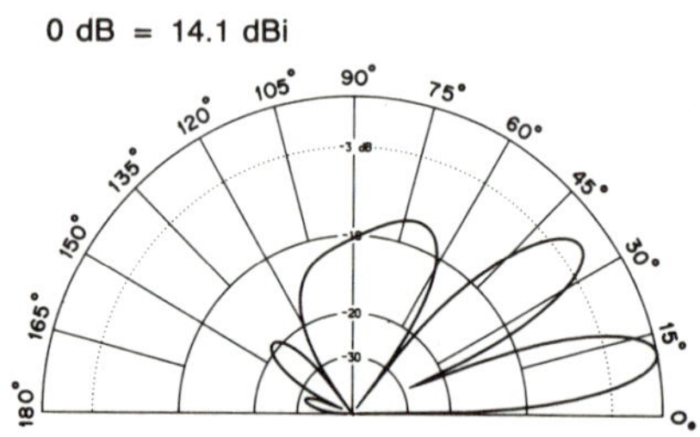

Fig. 5.2F — *H*-plane radiation pattern for a three-element Yagi 1.25 wavelengths above the ground.

0 dB = 14.2 dBi

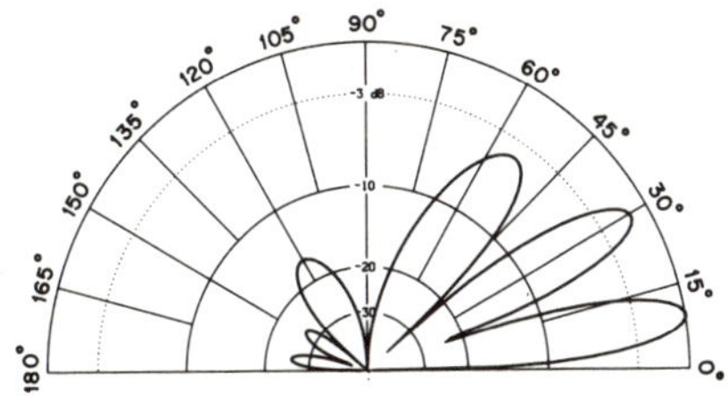

Fig. 5.2G — *H*-plane radiation pattern for a three-element Yagi 1.5 wavelengths above the ground.

0 dB = 14.2 dBi

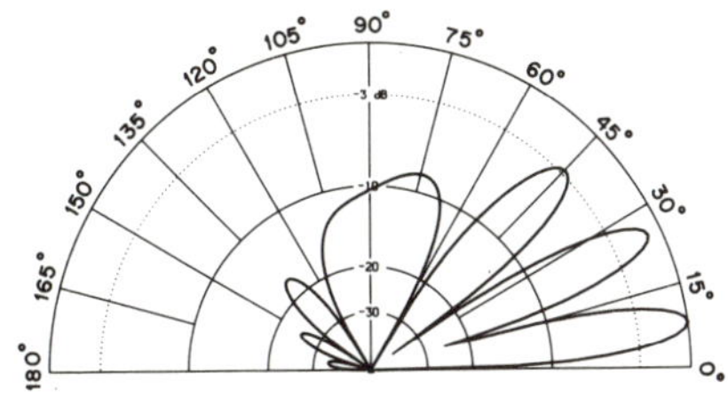

Fig. 5.2H — *H*-plane radiation pattern for a three-element Yagi 1.75 wavelengths above the ground.

0 dB = 14.3 dBi

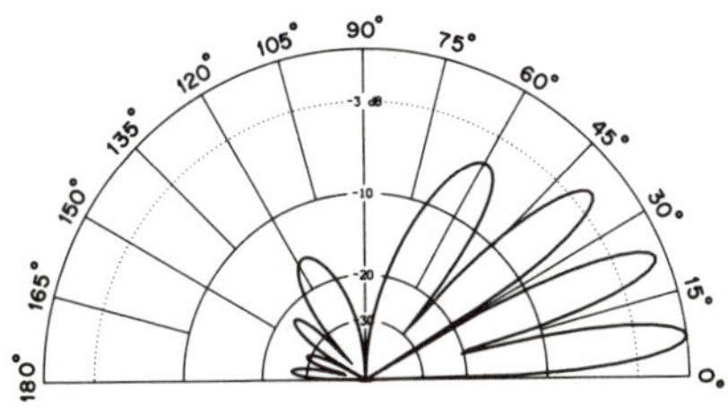

Fig. 5.2I — *H*-plane radiation pattern for a three-element Yagi 2.0 wavelengths above the ground.

0 dB = 14.3 dBi

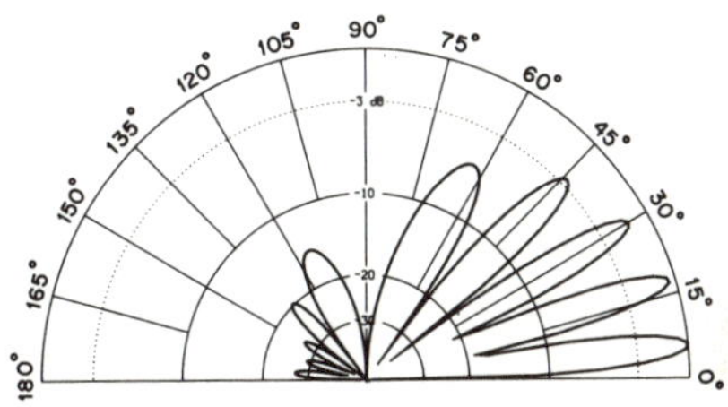

Fig. 5.2J — *H*-plane radiation pattern for a three-element Yagi 2.5 wavelengths above the ground.

0 dB = 14.3 dBi

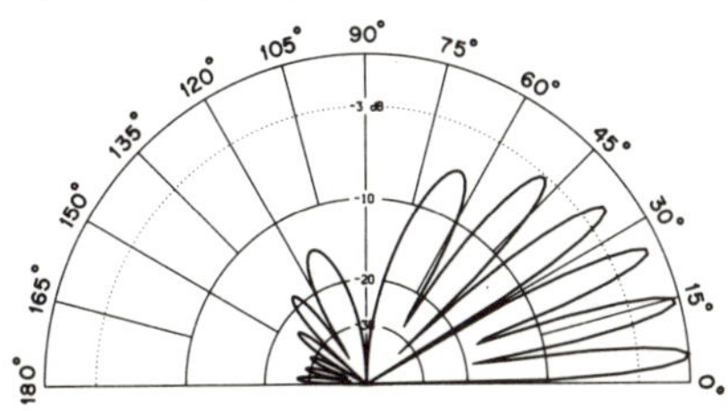

Fig. 5.2K — *H*-plane radiation pattern for a three-element Yagi 3.0 wavelengths above the ground.

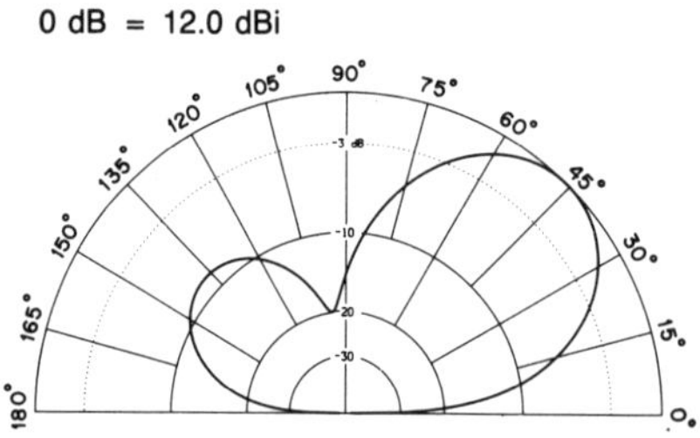

Fig. 5.3A — *H*-plane radiation pattern for a six-element Yagi 0.1 wavelengths above the ground.

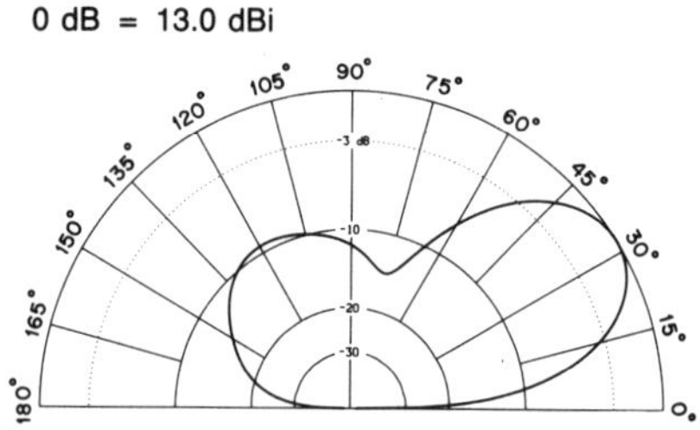

Fig. 5.3B — *H*-plane radiation pattern for a six-element Yagi 0.25 wavelengths above the ground.

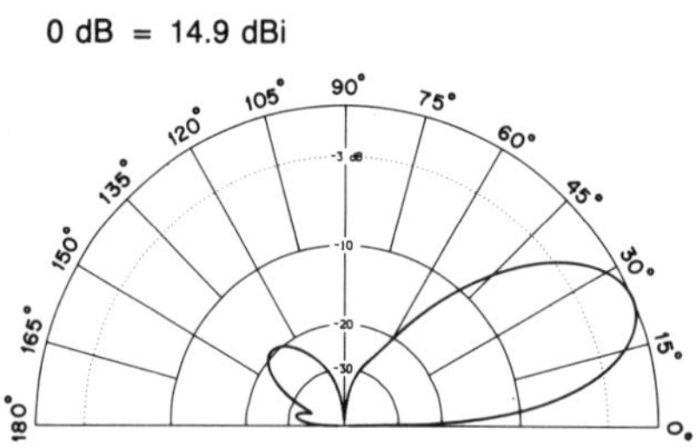

Fig. 5.3C — *H*-plane radiation pattern for a six-element Yagi 0.5 wavelengths above the ground.

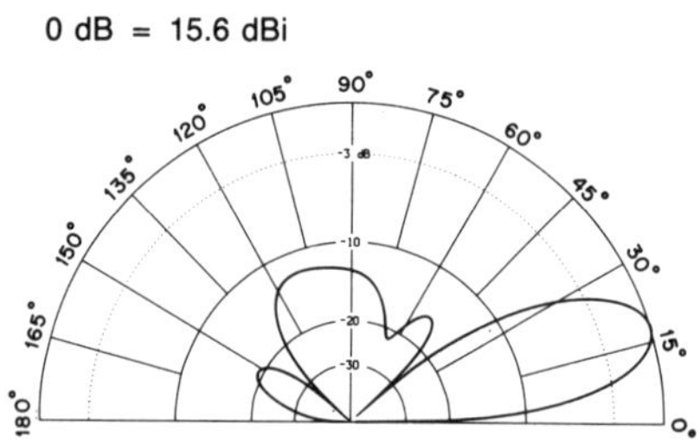

Fig. 5.3D — *H*-plane radiation pattern for a six-element Yagi 0.75 wavelengths above the ground.

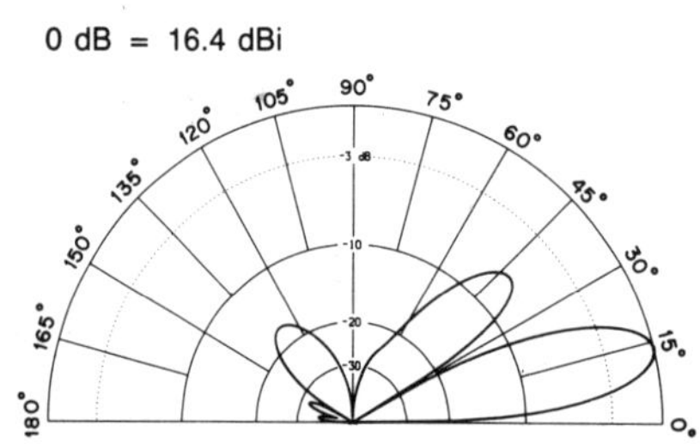

Fig. 5.3E — *H*-plane radiation pattern for a six-element Yagi 1.0 wavelengths above the ground.

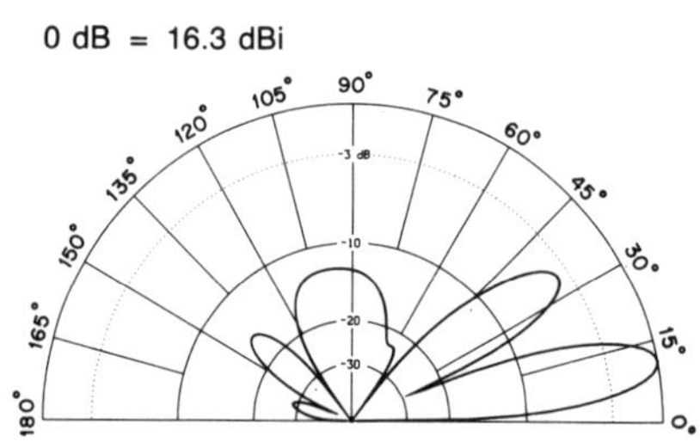

Fig. 5.3F — *H*-plane radiation pattern for a six-element Yagi 1.25 wavelengths above the ground.

0 dB = 16.6 dBi

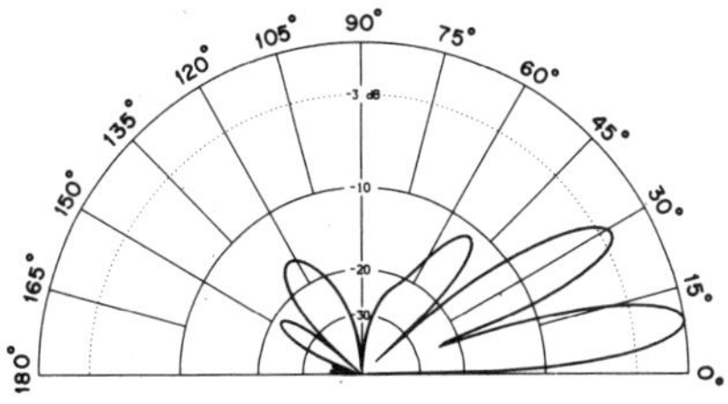

Fig. 5.3G — *H*-plane radiation pattern for a six-element Yagi 1.5 wavelengths above the ground.

0 dB = 16.5 dBi

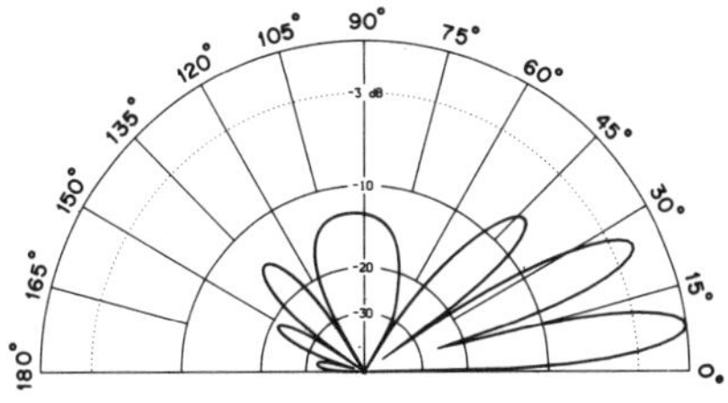

Fig. 5.3H — *H*-plane radiation pattern for a six-element Yagi 1.75 wavelengths above the ground.

0 dB = 16.7 dBi

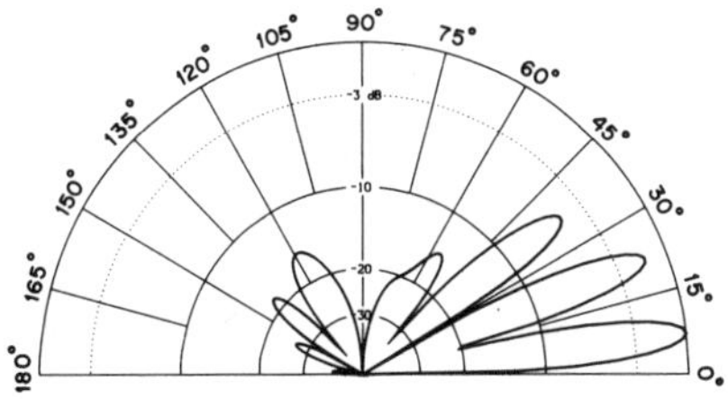

Fig. 5.3I — *H*-plane radiation pattern for a six-element Yagi 2.0 wavelengths above the ground.

0 dB = 16.7 dBi

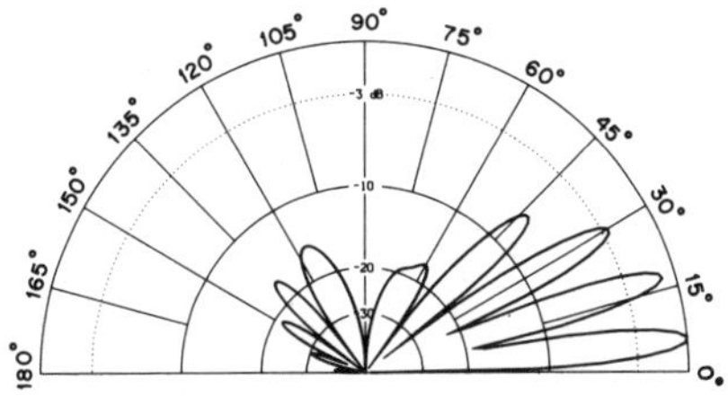

Fig. 5.3J — *H*-plane radiation pattern for a six-element Yagi 2.5 wavelengths above the ground.

0 dB = 16.7 dBi

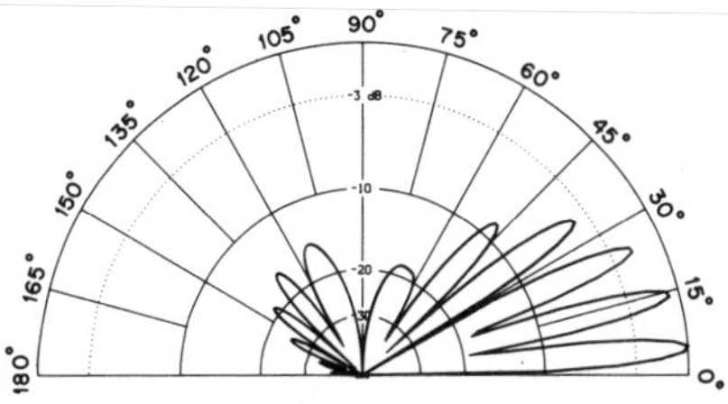

Fig. 5.3K — *H*-plane radiation pattern for a six-element Yagi 3.0 wavelengths above the ground.

the definition I use for *F/B* is the ratio of forward energy flux density at the best elevation angle to the reverse energy flux density at the same reverse elevation angle. Tables 5.2 and 5.3 show that this quantity fluctuates rather widely with antenna height; the cause of these fluctuations is the altered antenna element complex currents that result from the mutual coupling of antenna and its image. These mutual effects are large when the antenna is low and relatively small when the antenna is high. Note, however, that even when the antenna is three full wavelengths above ground, enough interaction occurs to noticeably alter the free space value.

Similarly, the antenna driving-point impedance fluctuates with antenna height. When the antenna is very low, for example, at a height of 0.1 wavelengths, driving-point resistance and reactance are far from their free space values. This shows dramatically that if one adjusts an antenna near the ground for best performance, it certainly will not be the best adjustment at final operating height.

These ground mutual effects, which alter the antenna element currents, are present to some degree at all antenna heights likely to be used in practice. This is tantamount to saying that the antenna over ground is not the same as the antenna in free space. An antenna optimized for free space will therefore not generally be quite optimum over ground. Obviously, one should really optimize the antenna over ground at the desired height.

Best Height

What is the best antenna height? Recall from Fig. 5.1 that one should strive for a large gain over a range of angles, for example, 3 to 17 degrees. An inspection of Figs. 5.2 and 5.3 shows that this occurs when the antenna height over ground is about 1.5 wavelengths. For 14 MHz radiation, this height would be about 100 feet. Practical operating experience does verify that such an antenna height gives excellent results. Note also that at a height of three wavelengths a deep lobe null occurs at an elevation angle of 10 degrees; this angle is sometimes important, such as for a great circle distance of 4500 km using two *F*-layer hops. Such a high antenna, even though excellent as a band opener at very low angles, would not be expected to be a good overall performer. I have tried a large 14 MHz antenna at a height of 2.6 wavelengths; from my location in New York State, the average European signals were found to be substantially inferior to those received from an antenna at a height of 1.5 wavelengths.

It is fortunate that an antenna at a height of 1.5 wavelengths over ground is not seriously degraded from its free space performance. Table 5.3 shows that the *F/B* ratio (at the design frequency) for the six-element beam is a superb 47 dB in free space and degrades only to a (still superb) value of 42 dB when the beam is mounted at 1.5 wavelengths above ground. In both cases, optimization procedures described in Chapter 3 can tune up the *F/B* ratio with only minor effects on other performance features.

I have carried out such an optimization by slightly varying the boom positions of the first and third directors; final beam specifications are shown in Table 5.4, and the performance around the frequency of best *F/B* is shown in Tables 5.5 and 5.6. Note that optimization requires very delicate boom position adjustments. These boom positions have been adjusted sufficiently well to give a *F/B* well over 100 dB at the central design frequency. It is interesting to note that different boom positions are needed for the free space Yagi and for the Yagi to be mounted over ground. This is because they are really slightly different antennas because of ground interaction.

Let me stress, as I did in Chapter 3, that although the Yagis are mathematically optimized to give a very high (but narrowband) *F/B* ratio, the basic model cannot really be trusted to this level of accuracy. It should

Table 5.4 — Element lengths and positions for six-element beams optimized by slight shifts in boom positions for D1 and D3. All element radii are 0.00052599 wavelengths.

	free space		*1.5 wavelengths high*	
element	*length* (λ)	*position* (λ)	*length* (λ)	*position* (λ)
reflector	0.49483	0.000	0.49483	0.000
driver	0.48073	0.150	0.48047	0.150
D1	0.44771	0.2983232	0.44771	0.303318
D2	0.44771	0.450	0.44771	0.450
D3	0.44771	0.6015036	0.44771	0.597591
D4	0.44771	0.750	0.44771	0.750

Table 5.5 — Characteristics of the six-element Yagi optimized in free space.

normalized frequency	*gain (dBi)*	*F/B (dB)*	*angle (degrees)*	*resistance (ohms)*	*reactance (ohms)*
0.996	10.58	26.6	0	22.0	−6.2
0.998	10.64	32.7	0	21.6	−3.1
1.000	10.70	134.8	0	21.3	−0.1
1.002	10.75	32.7	0	21.0	3.1
1.004	10.80	26.7	0	20.8	6.3

Table 5.6 — Characteristics of the six-element Yagi optimized at 1.5 wavelengths above ground.

normalized frequency	*gain (dBi)*	*F/B (dB)*	*angle (degrees)*	*resistance (ohms)*	*reactance (ohms)*
0.996	16.50	26.5	9	21.9	−5.4
0.998	16.55	32.5	9	21.6	−2.4
1.000	16.61	119.8	9	21.4	0.7
1.002	16.65	32.6	9	21.2	3.8
1.004	16.69	26.6	9	21.0	7.0

be quite possible in principle to carry out this type of optimization experimentally on a real Yagi; the basic behavior should be similar, but the final boom positions might be slightly different.

Antenna Upward Tilt

Note that when a Yagi is mounted over ground the lowest lobe of radiation has a maximum at an elevation angle usually sufficiently high that the direct wave from the antenna is somewhat reduced from its peak free space value. It is interesting to see if any improvement could be made by purposely tipping the antenna boom upwards to increase the direct wave. Unfortunately, tipping the antenna upward automatically tips the image downward by the same angle; the net result is that, while the direct wave is increased, the reflected wave is decreased, and by a greater amount.

Table 5.7 — Characteristics of a representative six-element Yagi one wavelength above ground with the boom tipped upward.

Fig.	*tip angle (degrees)*	*gain (dBi)*	*F/B (dB)*	*angle (degrees)*	*resistance (ohms)*	*reactance (ohms)*
5.3E	0	16.36	37.7	14	21.9	−0.4
5.4A	5	16.31	33.7	14	22.0	−0.6
5.4B	10	16.12	29.9	14	22.0	−0.7
5.4C	15	15.78	26.8	13	21.8	−0.9
5.4D	20	15.26	23.8	13	21.6	−0.9

As an example, consider the six-element Yagi mounted one wavelength above ground. The maximum gain of 16.36 dBi occurs at an angle of 14 degrees as shown in Table 5.3. Table 5.7 and Fig. 5.4 show the result if the antenna is tipped upward at angles of 5, 10, 15, and 20 degrees. It is

0 dB = 16.3 dBi

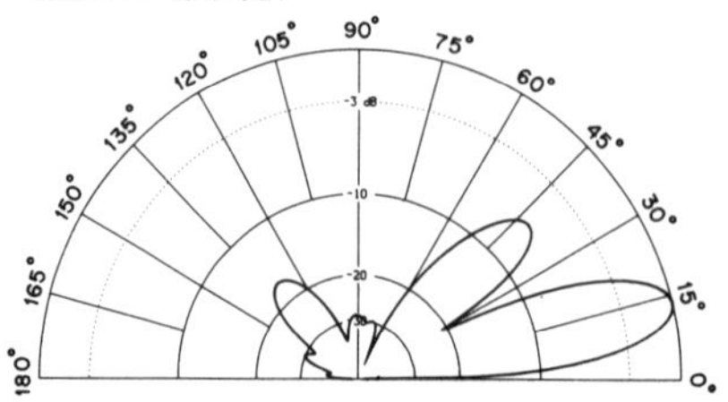

Fig. 5.4A — *H*-plane radiation patterns for a six-element beam at a height of one wavelength with boom tilted upward 5 degrees.

0 dB = 16.1 dBi

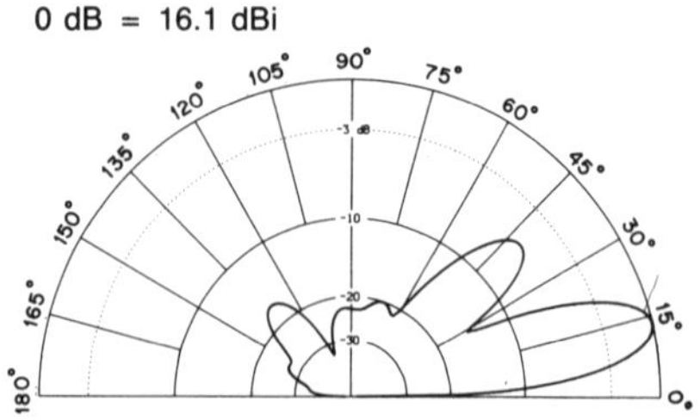

Fig. 5.4B — *H*-plane radiation patterns for a six-element beam at a height of one wavelength with the boom tilted upward 10 degrees.

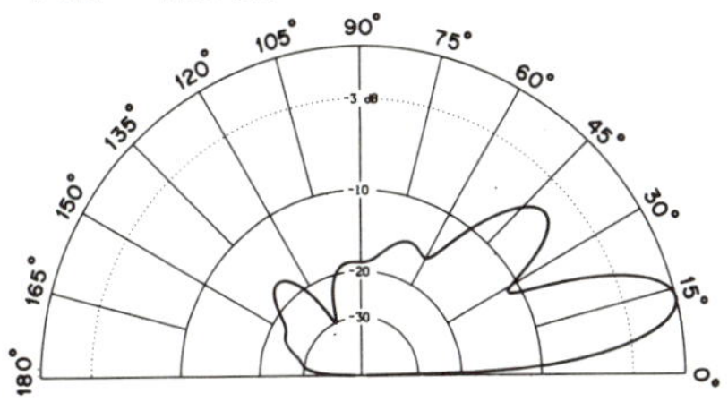

Fig. 5.4C — *H*-plane radiation patterns for a six-element beam at a height of one wavelength with the boom tilted upward 15 degrees.

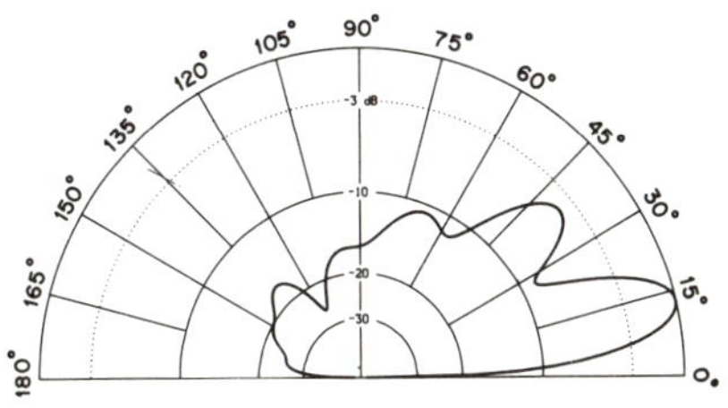

Fig. 5.4D — *H*-plane radiation patterns for a six-element beam at a height of one wavelength with the boom tilted upward 20 degrees.

easy to see that maximum overall gain is actually best when the antenna is parallel to the ground plane; as one tips the antenna the peak lobe gain is reduced slightly and the deep nulls between lobes tend to become shallower. This is precisely the behavior expected from a consideration of the vectorial addition of direct and reflected waves.

Summary

1) Although ground is difficult to characterize, there is reason to believe that for horizontal polarization a good model is an ideal, infinitely conducting plane.

2) The *H*-plane (vertical) pattern consists of a number of lobes caused by the interference of direct and ground-reflected waves. The first (lowest) lobe is the strongest; succeeding lobes are reduced somewhat in gain by the natural free space directivity of the antenna.

3) Mutual effects between the antenna and ground cause antenna element currents to change; these changes cause significant alterations to the antenna properties. The most noticeable variations occur in *F/B* ratio, but there are also significant variations in gain and driving-point impedance.

4) Best overall antenna performance appears to occur if the antenna height is about 1.5 wavelengths. This is not a critical figure, but it is believed that three wavelengths is probably too high.

5) Adjusting an antenna near the ground for best performance guarantees that the antenna performance will not be optimum at the operating height. Because of the significant mutual effects with the ground, the antenna should be optimized at its final operational height. Generally, this optimization will not be quite the same as the optimized free space antenna.

6) Large antennas are more handicapped at low heights than are small

antennas; this is due to their higher natural free space directivity.

7) For best gain the antenna boom should, in principle be parallel to the effective ground plane surface; however, the degree of parallelism required is not critical. Tipping the antenna upward to improve gain will, in actual fact, decrease maximum gain.

NOTES:

[1]*The ARRL Antenna Book,* 13th Edition, American Radio Relay League, 1974.

[2]Kift, F. "The propagation of High-Frequency Radio Waves to Long Distances," *Proceedings of IEE*, 107 Part B, March 1960, page 127.

CHAPTER 6

STACKING

This chapter addresses the use of multiple Yagi antennas arranged into a coherent antenna system. The number of potential arrangements is unlimited, but certain basic configurations deserve detailed analysis because they have attractive properties. To start, I shall limit the discussion to systems where the individual Yagi antennas are all physically identical and aligned for maximum radiation in the same direction. Moreover, to ensure that each Yagi contributes to the overall main radiated wave front in a coherent manner, I shall limit the configurations to those in which the Yagi positions (say, for example, the reflector end of the boom) lie in a plane perpendicular to boom direction. Usually all of the Yagis are coherently excited by the same driver current (magnitude and phase). Using identical Yagis positioned in such a plane helps maintain a uniform radiated pattern over a desired frequency band. The overall system beam pattern can be pointed in azimuth only by mechanically rotating the entire system. The radiated beam from a mechanically fixed (system) array of laterally spaced Yagi antennas can, in principle, be steered by changing the excitation phase to each Yagi antenna. However, the beam quality generally deteriorates. Such mechanically fixed, electrically steered phased arrays are not considered here.

The overall system array can be viewed as a large-area aperture illuminated in a quasi-uniform way by the individual Yagi antennas. So long as the individual Yagi antennas are not too far apart (so that illumination is relatively uniform), the system gain should be proportional to the total effective aperture area. The system beam pattern should also show an angular width inversely proportional to the aperture dimension. Thus, in concept, a horizontal array of Yagi antennas (horizontally polarized) should produce a narrow horizontal system beam pattern; similarly, a vertical array of Yagi antennas (horizontally polarized) should produce a narrow vertical system beam pattern.

We must consider the system array over ground; in this case all of the effects mentioned previously (Chapter 5) will occur. Recall that ionospheric paths over the earth primarily favor low radiation angles (up to, say, 20 degrees); moreover, this whole range of antenna radiation angles should be covered to accommodate a continuous range of great circle distances, as well as different multi-mode ionospheric paths. We shall see that, by vertically stacking two or more horizontally polarized Yagis over ground, it is possible to improve significantly low-angle performance (over that of a single Yagi antenna over ground) without reducing the azimuthal coverage. This improved result comes about through a suppression of otherwise useless radiation at the higher angles.

Vertical Stacking Arrangements

For Amateur Radio communications, relatively wide horizontal or azimuthal coverage is generally desirable, not only to make a given contact less sensitive to critical beam heading but to accommodate the many occasions in which the communication path is somewhat skewed due to ionospheric conditions. Wide azimuthal coverage is especially desirable under contest conditions, where it is advantageous to have the beam simultaneously illuminate the largest desired Amateur population. So, a horizontal array of Yagi antennas doesn't appear as desirable as a vertical stack; therefore I shall not attempt analyses of such horizontal arrays.

Vertically stacked Yagi arrays are now in reasonably wide use; these require a supporting mast. If the stacking separation is large (which we shall find desirable), the large mast must be entirely rotatable and, of course, very rugged mechanically. Such a mast, including its foundation, is a major undertaking.

Two interesting variations of this system are not as formidable. The first variation is a stacked Yagi antenna array offset from a fixed tower. The offset allows simultaneous rotation of the Yagi antennas over a range in azimuth of about 300 degrees; at either end of this range, the antennas are designed to nest around the mast. I use this construction for a 28 MHz stacked six-element Yagi antenna system on a Rohn 45 guyed mast. It works very well and is cost effective.

The second variation is to use a fixed mast with the top Yagi antenna fully rotatable and a second, lower, Yagi antenna fixed in a preferred direction. This is a particularly interesting variation for contest operation, especially on the lower frequencies where the mast must be very high. My 7 MHz system is a good example. A full-sized, three-element beam is fully rotatable on top of a 180 foot Rohn 45 guyed mast. A second full-sized, three-element beam is fixed at 90 feet, which is aimed at Europe. Thus, in the European direction, full stacking is available; in all other directions the top beam can be used alone. Moreover, it is easy to excite both beams and activate two azimuthal directions simultaneously, or it is also possible to switch instantly from one direction to another without losing the normal time to turn the large Yagi antenna. I have found the flexibility of this system to be very helpful in many situations.

Excitation

For all types of stacked arrays, I have found it useful to provide a switching system that allows operation of each Yagi independently or both together. When only high-angle radiation is desired, the lower antenna is usually best. For lower angles of radiation, the combined stack is better. It is easy to arrange such a switch using conventional relays and quarter-wave coaxial transformers; a practical system is shown in Fig. 6.1 for two stacked Yagi antennas.

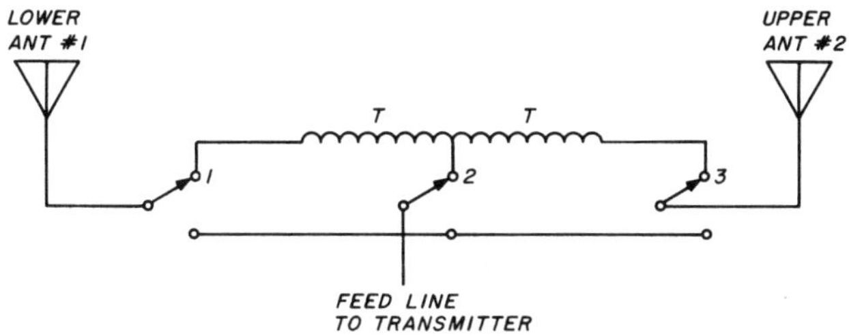

CONTACTS FOR RELAYS 1, 2, 3 SHOWN IN NORMAL (UNEXCITED) STATE

RELAY	#1,2 BOTH	#1 LOWER	#2 UPPER
1	OFF	ON	OFF
2	OFF	ON	ON
3	OFF	OFF	ON

RELAY EXCITATION SCHEDULE

Fig. 6.1 — Both, lower, or upper (BLU) switching system. *T* is a quarter-wave transformer of 70 ohm coaxial cable.

The relays may have to be compensated by small shunt capacitors if their series inductance is too large. The relay box should be mounted on the mast about half way between the Yagi antennas. Extension to more than two stacked Yagi antennas is equivalently easy. However, the particular scheme will depend on the way in which power is to be split between all Yagi antennas.

Because of these various excitation techniques, it is desirable to compute not only the properties of a vertically stacked Yagi system, but the properties of the individually excited Yagi antennas. Two complicating problems arise. First, not only is a single Yagi antenna over ideal ground not the same antenna as in free space, but it is further changed by all other Yagi antennas as well as their ground images. This is true even if all other Yagi antennas are not driven. To some extent their elements will be parasitically excited by the single driven Yagi antenna. This means that the computation for a single Yagi must be carefully made to account fully for all the parasites and images in its local field.

Second, if only the top Yagi is rotatable, the performance of the single lower antenna alone will depend on the relative azimuthal orientation of the two antennas. In this case it is instructive to compute three cases: parallel, orthogonal, and antiparallel orientations.

Two-Array Stack

Let us now choose some representative horizontally polarized stacking arrangements over flat, ideal, ground and compute their theoretical performance. I shall present computed *H*-plane patterns over the range of elevation angles of interest. The *E*-plane pattern over ideal ground is, of course,

Table 6.1 — Performance of a two-array stack of three-element Yagis for different combinations of heights.

Fig. 6.2	*height lower (λ)*	*height upper (λ)*	*both gain (dBi)*	*both angle (deg)*	*both F/B (dB)*	*lower gain (dBi)*	*lower angle (deg)*	*lower F/B (dB)*
A	0.30	0.75	14.30	21	21.4	11.41	41	19.0
B	0.375	0.75	14.38	21	20.0	11.69	36	27.4
C	0.45	0.75	14.41	21	18.8	12.01	31	23.4
D	0.60	1.50	15.55	11	18.2	13.80	24	21.0
E	0.75	1.50	16.27	11	13.9	13.80	20	22.6
F	0.90	1.50	16.58	11	19.4	13.72	16	25.4
G	0.90	2.25	15.41	8	23.8	13.83	15	37.5
H	1.125	2.25	16.06	8	28.7	14.10	13	25.3
I	1.35	2.25	16.74	7	19.8	14.02	11	25.8
J	1.00	3.00	14.97	6	30.7	14.09	14	23.3
K	1.50	3.00	16.11	6	22.9	14.18	9	25.8
L	2.00	3.00	16.86	5	32.9	14.19	7	36.0

Fig. 6.2	*height lower (λ)*	*height upper (λ)*	*upper gain (dBi)*	*upper angle (deg)*	*upper F/B (dB)*	*lower alone gain (dBi)*	*lower alone angle (deg)*	*lower alone F/B (dB)*
A	0.30	0.75	13.54	18	29.0	11.54	37	22.9
B	0.375	0.75	13.39	18	23.7	12.20	32	47.8
C	0.45	0.75	13.32	18	20.1	12.82	29	28.4
D	0.60	1.50	13.96	9	23.9	13.67	23	19.5
E	0.75	1.50	13.71	9	16.0	13.84	19	25.3
F	0.90	1.50	13.83	9	18.9	13.94	16	31.2
G	0.90	2.25	14.31	6	30.8	13.94	16	31.2
H	1.125	2.25	14.19	6	29.9	14.08	13	24.2
I	1.35	2.25	14.00	6	21.3	14.13	11	37.2
J	1.00	3.00	14.22	5	30.8	14.04	14	25.5
K	1.50	3.00	14.31	5	24.6	14.17	9	26.5
L	2.00	3.00	14.20	5	34.1	14.24	7	27.3

zero everywhere. The plots show not only how well the overall system performs at the important low angles, but also what may be sacrificed at the higher angles, which are occasionally useful. Two basic Yagi designs are used. They are the same three-element beam (boom = 0.3 wavelengths) and the same six-element beam (boom = 0.75 wavelengths) shown in Table 5.1.

I shall start with two stacked, identical beams over ground. In practice, the height of the upper beam will be fixed at the overall mast height. The placement of the lower antenna will be made at some lower position. It is interesting to understand the tradeoffs involved in the height of the lower antenna. I shall choose, for illustrative purposes, four different heights for the upper beam, and for each of these cases, three different heights for the lower beam. All heights are expressed in wavelengths at the central design frequency. Tables 6.1 and 6.2 show computed results for all these cases. These tables also refer to Figs. 6.2 and 6.3, which display detailed *H*-plane patterns for all cases.

Table 6.2 — Performance of a two-array stack of six-element Yagis for different combinations of heights.

Fig. 6.3	*height lower (λ)*	*upper (λ)*	*both gain (dBi)*	*angle (deg)*	*F/B (dB)*	*lower gain (dBi)*	*angle (deg)*	*F/B (dB)*
A	0.30	0.75	15.52	20	23.2	12.78	33	15.8
B	0.375	0.75	15.60	20	20.2	13.06	27	26.1
C	0.45	0.75	15.61	19	18.2	13.22	24	26.2
D	0.60	1.50	17.47	11	22.5	15.11	21	21.6
E	0.75	1.50	17.28	11	14.0	15.10	19	16.7
F	0.90	1.50	18.10	11	23.3	15.73	16	38.8
G	0.90	2.25	18.00	8	31.2	16.06	15	24.7
H	1.125	2.25	18.42	8	29.2	16.31	12	49.0
I	1.35	2.25	18.60	7	19.9	16.22	11	22.6
J	1.00	3.00	17.34	6	39.7	16.38	14	35.5
K	1.50	3.00	18.71	6	32.9	16.61	9	39.1
L	2.00	3.00	19.24	5	25.5	16.59	7	27.1

Fig. 6.3	*height lower (λ)*	*upper (λ)*	*upper gain (dBi)*	*angle (deg)*	*F/B (dB)*	*lower alone gain (dBi)*	*angle (deg)*	*F/B (dB)*
A	0.30	0.75	15.56	17	21.7	13.39	30	17.3
B	0.375	0.75	15.22	17	17.6	13.96	27	24.0
C	0.45	0.75	14.75	17	14.6	14.51	25	30.0
D	0.60	1.50	16.40	9	30.3	15.18	21	23.7
E	0.75	1.50	16.06	9	17.8	15.60	18	21.5
F	0.90	1.50	16.08	9	24.6	16.16	15	29.1
G	0.90	2.25	15.16	6	57.5	16.16	15	29.1
H	1.125	2.25	16.55	6	31.3	16.32	12	33.5
I	1.35	2.25	16.24	6	21.7	16.38	10	32.3
J	1.00	3.00	16.68	5	41.2	16.36	14	37.7
K	1.50	3.00	16.76	5	36.5	16.59	9	41.7
L	2.00	3.00	16.58	5	26.2	16.68	7	45.2

Note that each figure has several curves: one for the combined stacked performance (solid line); one for the lower antenna alone (dotted line); one for the upper antenna alone (dashed line); and, where applicable, what the lower antenna only would show if no upper antenna were physically present (broken solid line). In the second and third cases, both antennas are physically present, but only one is driven (all nondriven elements act as parasites).

An examination of Tables 6.1 and 6.2, and especially the *H*-plane patterns of Figs. 6.2 and 6.3, reveals a number of interesting and important characteristics of these simple, vertically stacked systems. Table 6.1 shows the maximum gain and corresponding elevation angle for each case of a stacked pair of three-element beams. Also shown is the *F/B* ratio, which we know varies with the exact element complex currents, which in turn are influenced by the mutual impedances to all other elements. Table 6.2 shows the equivalent quantitites for the stacked pair of six-element beams.

Note from these tables that the smaller values of overall antenna mast

combination	curve
both	solid
lower	dots
upper	dashes
lower alone	chain-dots

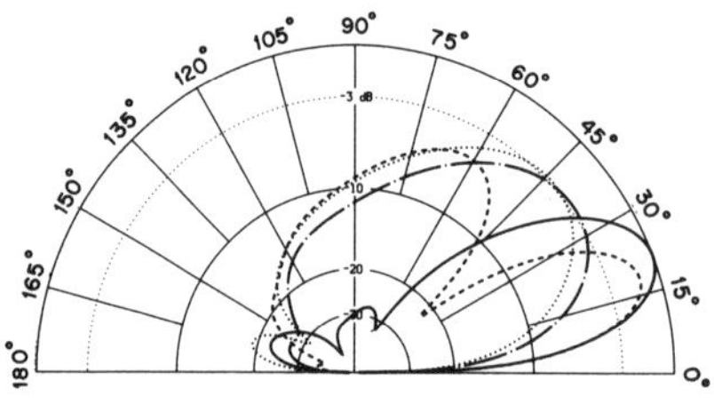

Fig. 6.2A — *H*-plane radiation pattern for two three-element Yagis at 0.3 and 0.75 wavelengths above ground.

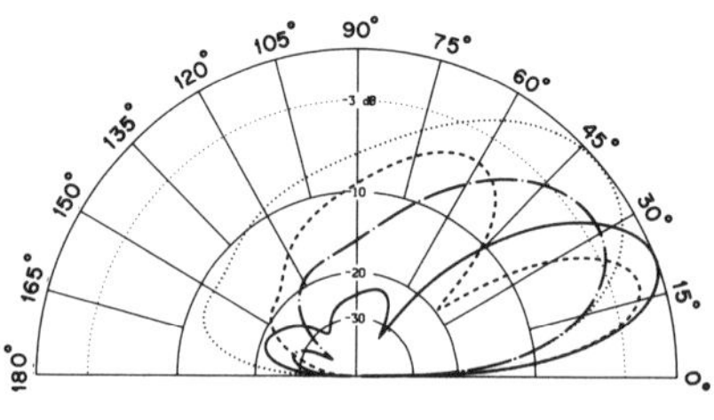

Fig. 6.2B — *H*-plane radiation pattern for two three-element Yagis at 0.375 and 0.75 wavelengths above ground.

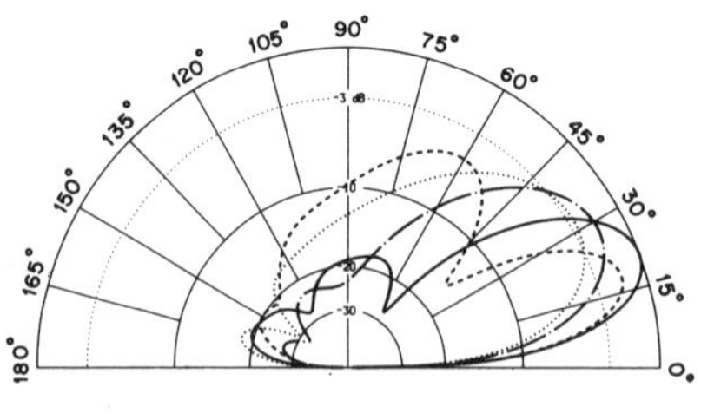

Fig. 6.2C — *H*-plane radiation pattern for two three-element Yagis at 0.45 and 0.75 wavelengths above ground.

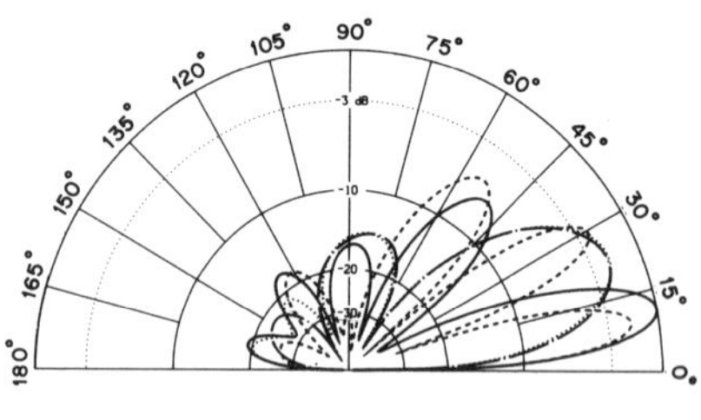

Fig. 6.2D — *H*-plane radiation pattern for two three-element Yagis at 0.6 and 1.5 wavelengths above ground.

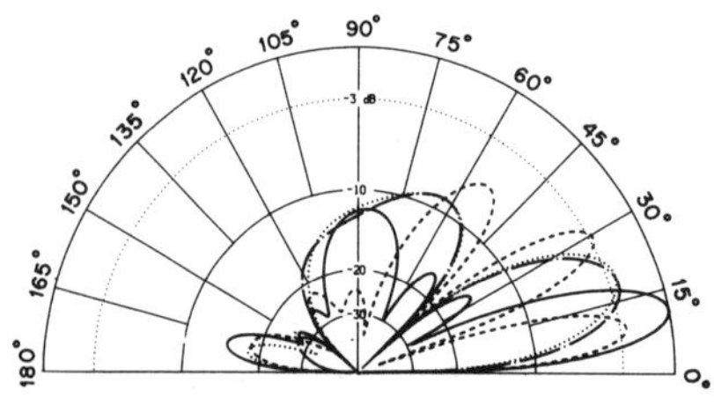

Fig. 6.2E — *H*-plane radiation pattern for two three-element Yagis at 0.75 and 1.5 wavelengths above ground.

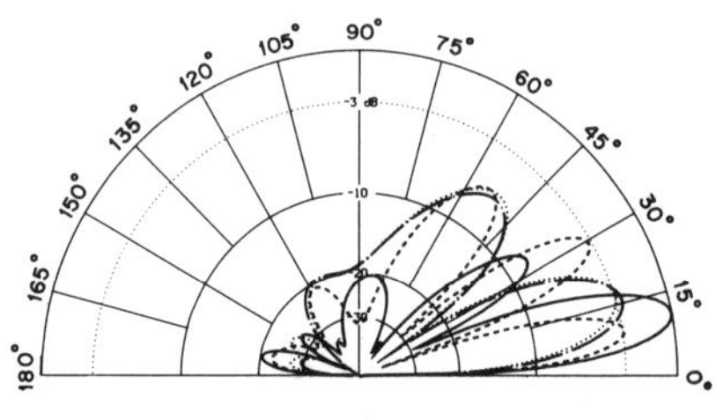

Fig. 6.2F — *H*-plane radiation pattern for two three-element Yagis at 0.9 and 1.5 wavelengths above ground.

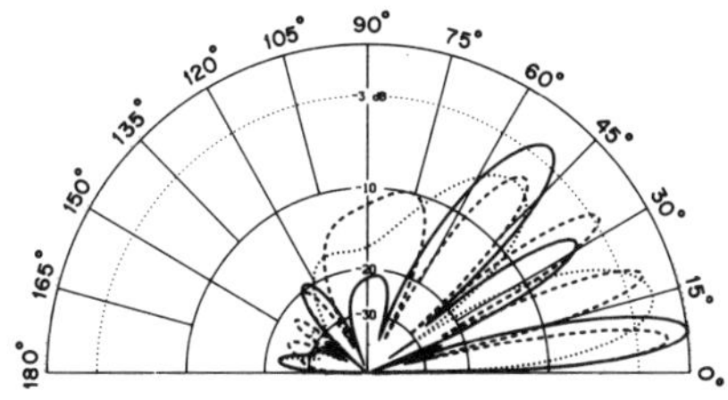

Fig. 6.2G — *H*-plane radiation pattern for two three-element Yagis at 0.9 and 2.25 wavelengths above ground.

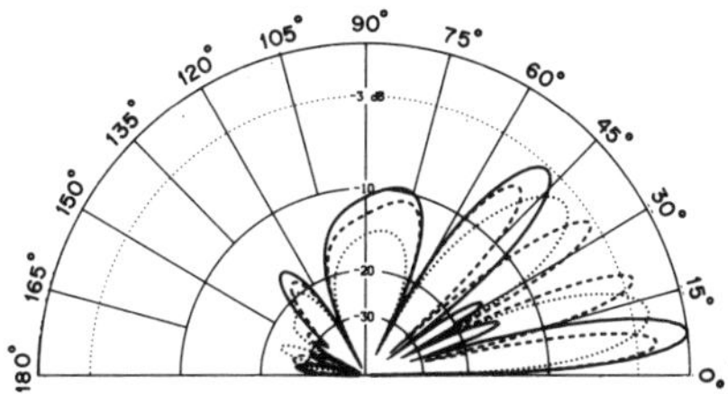

Fig. 6.2H — *H*-plane radiation pattern for two three-element Yagis at 1.125 and 2.25 wavelengths above ground.

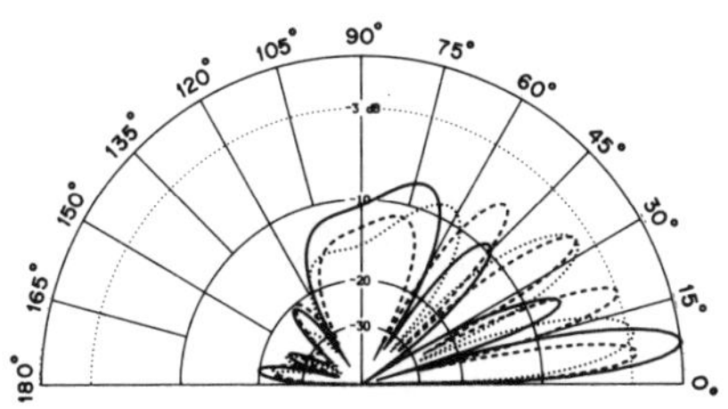

Fig. 6.2I — *H*-plane radiation pattern for two three-element Yagis at 1.35 and 2.25 wavelengths above ground.

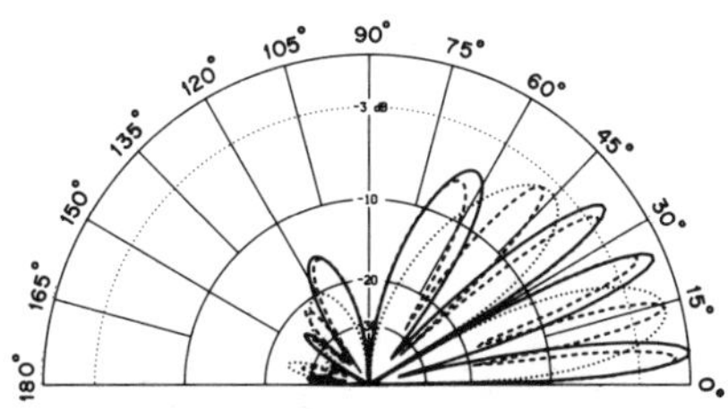

Fig. 6.2J — *H*-plane radiation pattern for two three-element Yagis at 1.0 and 3.0 wavelengths above ground.

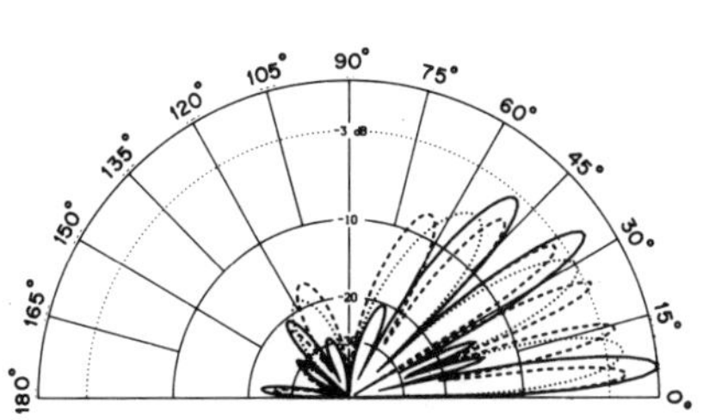

Fig. 6.2K — *H*-plane radiation pattern for two three-element Yagis at 1.5 and 3.0 wavelengths above ground.

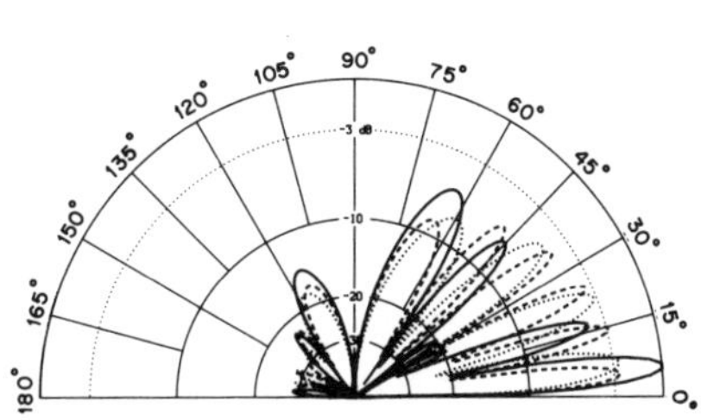

Fig. 6.2L — *H*-plane radiation pattern for two three-element Yagis at 2.0 and 3.0 wavelengths above ground.

combination	curve
both	solid
lower	dots
upper	dashes
lower alone	chain-dots

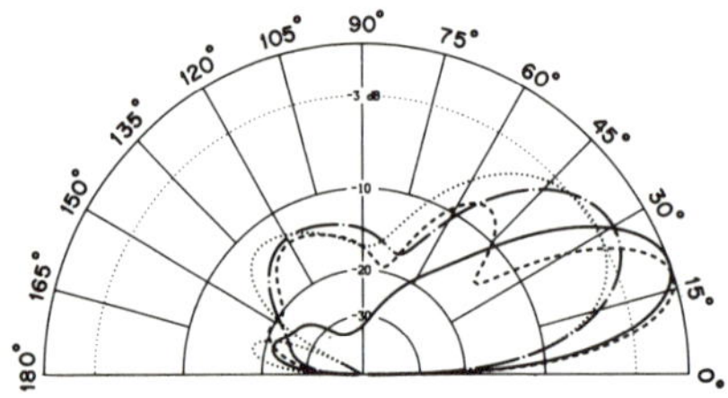

Fig. 6.3A — *H*-plane radiation pattern for two six-element Yagis at 0.3 and 0.75 wavelengths above ground.

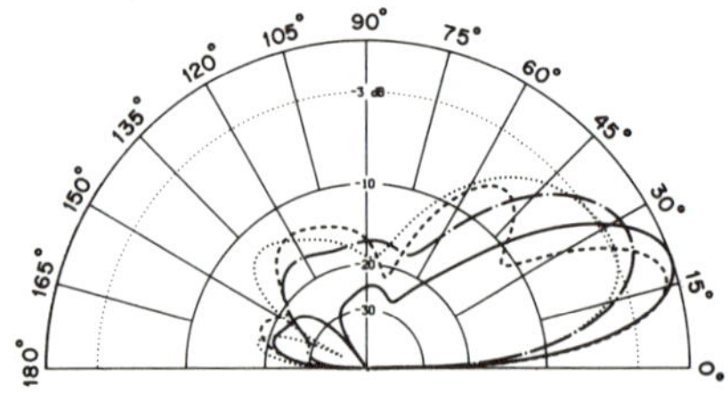

Fig. 6.3B — *H*-plane radiation pattern for two six-element Yagis at 0.375 and 0.75 wavelengths above ground.

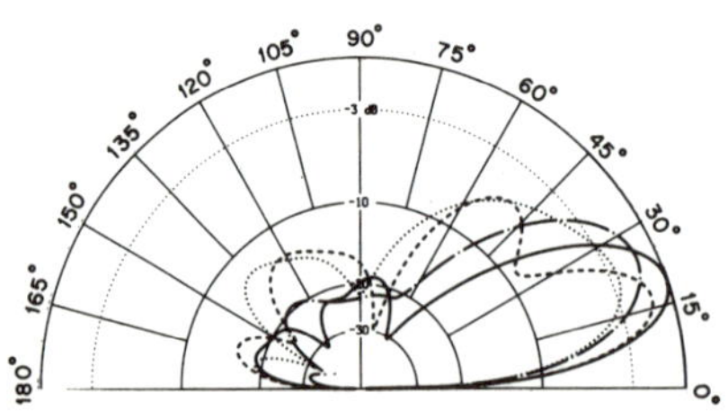

Fig. 6.3C — *H*-plane radiation pattern for two six-element Yagis at 0.45 and 0.75 wavelengths above ground.

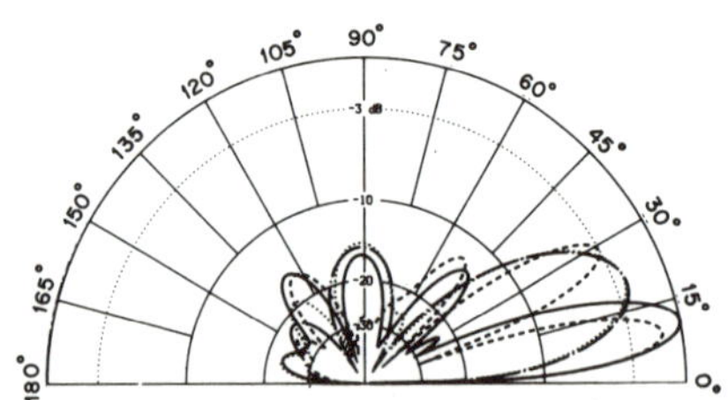

Fig. 6.3D — *H*-plane radiation pattern for two six-element Yagis at 0.6 and 1.5 wavelengths above ground.

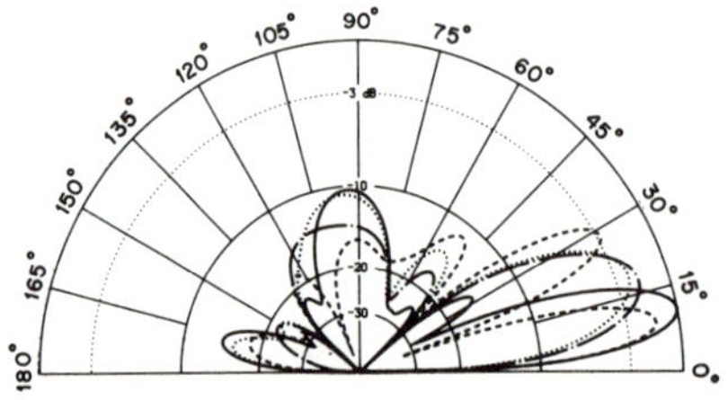

Fig. 6.3E — *H*-plane radiation pattern for two six-element Yagis at 0.75 and 1.5 wavelengths above ground.

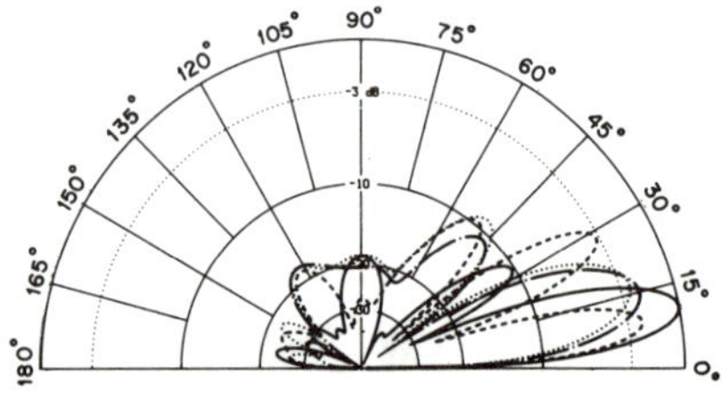

Fig. 6.3F — *H*-plane radiation pattern for two six-element Yagis at 0.9 and 1.5 wavelengths above ground.

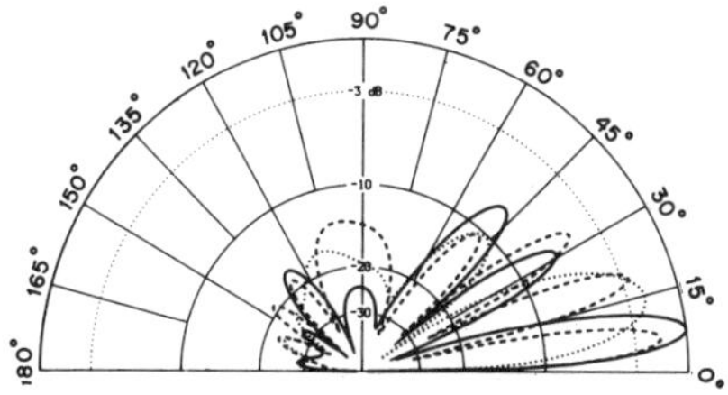

Fig. 6.3G — *H*-plane radiation pattern for two six-element Yagis at 0.9 and 2.25 wavelengths above ground.

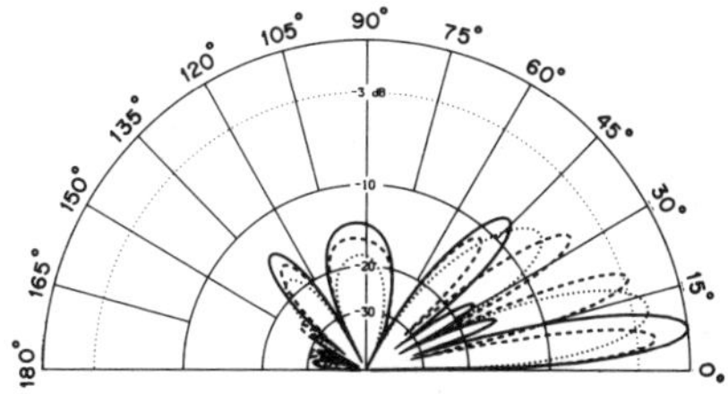

Fig. 6.3H — *H*-plane radiation pattern for two six-element Yagis at 1.125 and 2.25 wavelengths above ground.

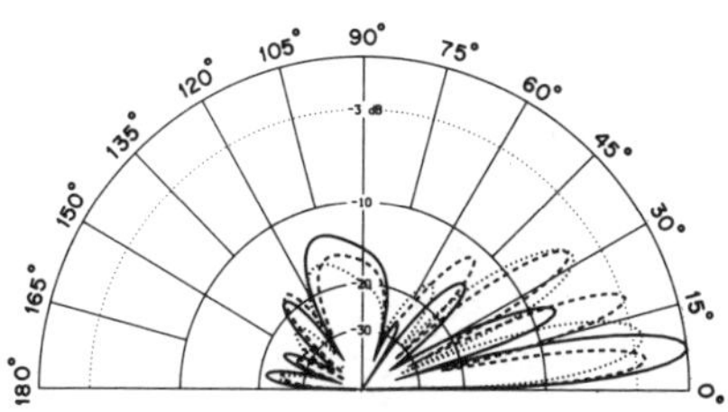

Fig. 6.3I — *H*-plane radiation pattern for two six-element Yagis at 1.35 and 2.25 wavelengths above ground.

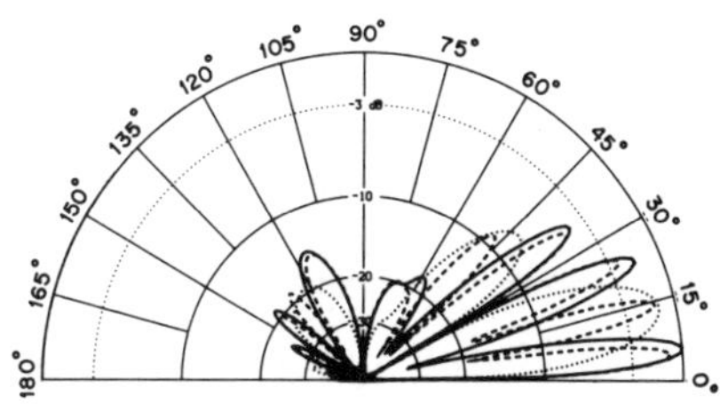

Fig. 6.3J — *H*-plane radiation pattern for two six-element Yagis at 1.0 and 3.0 wavelengths above ground.

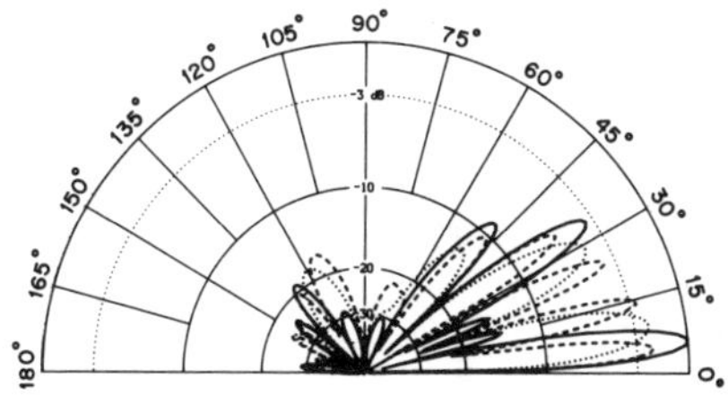

Fig. 6.3K — *H*-plane radiation pattern for two six-element Yagis at 1.5 and 3.0 wavelengths above ground.

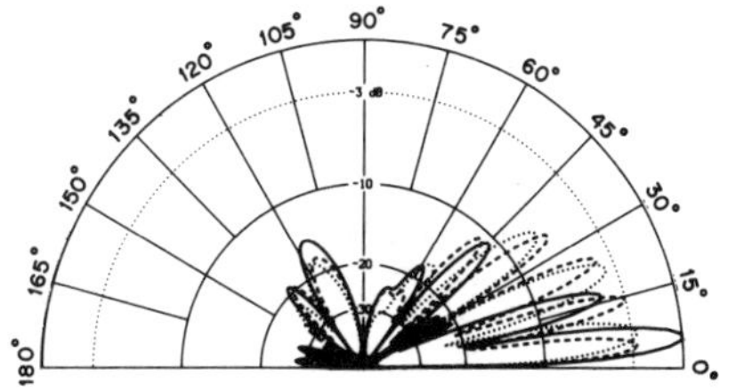

Fig. 6.3L — *H*-plane radiation pattern for two six-element Yagis at 2.0 and 3.0 wavelengths above ground.

height do not give as much overall maximum gain as the higher antennas; this gain deficit is more severe for the six-element beams than for the three-element beams. This is the same general result previously obtained for single antennas over ground (Chapter 5); it results from the same phenomenon; that is, the natural increased free space directivity of the larger Yagi antennas reduces the gain potential at the higher elevation angles required for the lower antennas.

Note also from these tables that the exact placement of the lower Yagi antenna does not markedly influence the stacked maximum gain of the system but usually does significantly affect the angle of the lower antenna radiation. Note also that the excellent free space *F/B* ratio can be significantly affected by stacking; it is most strongly affected when the stack spacing is small and where the number of (adjacent) parasites is large, for example, especially the first three cases in Table 6.2.

To properly assess all of these stacked Yagi antenna systems, it is necessary to look at the *H*-plane (elevation angle) patterns shown in Figs. 6.2 and 6.3. It is instantly clear that excellent stacked coverage (solid line) of the crucially important zero to 20 degree elevation angles requires a reasonably high system (more than one wavelength) but not too high (less than 2.5 wavelengths). Above the first main lobe of radiation the patterns are quite varied; it is helpful to understand the basic reasons for these variations. Fig. 6.4 shows a simplified sketch of the two Yagi antennas above ground, each one represented on this diagram by a point. The lower antenna is at height h_L and the upper one is at height h_U; also shown are the image antennas below ground at heights of $-h_L$ and $-h_U$, respectively. Note that at an elevation angle, ϕ, the radiation from the lower antenna lags that from the upper antenna by a distance $(h_U - h_L)\sin\phi$. This phase lag causes the pair of antennas to interfere both constructively and destructively. At certain values of ϕ, which I shall designate ϕ_P, destructive interference will be complete and produce a radiation pattern null. Since the phase lag between the two antennas above ground is identical to that between the two images below ground, the overall radiation will also show these nulls where

$$\phi_P = \sin^{-1}[(n + ½)/(h_U - h_L)] \qquad (6.1)$$

where n can take on integer values starting with zero (0, 1, 2, . . .).

Now, from Fig. 6.4, note that the radiation from the image pair (which is excited out of phase with the real antenna pair) further lags by a distance $(h_U + h_L)\sin\phi$. Thus nulls will also occur in the overall pattern due to ground reflections at values of ϕ which I shall designate as ϕ_G where:

$$\phi_G = \sin^{-1}[m/(h_U + h_L)] \qquad (6.2)$$

where m can assume integer values (0, 1, 2, ...). As an example, consider Fig. 6.3L where h_U is three wavelengths and h_L is two wavelengths. Equa-

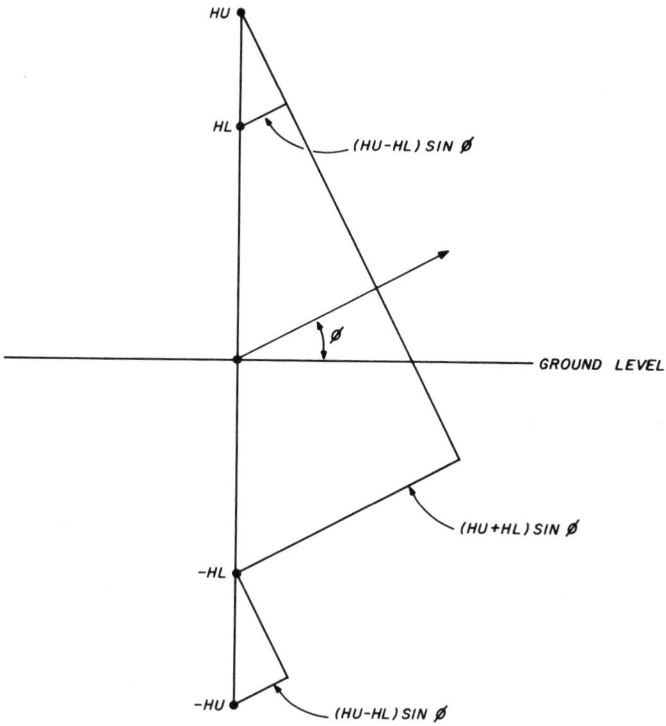

Fig. 6.4 — Diagram of two-array stack over ground, showing phase lags for each antenna and its image.

tions (6.1) and (6.2) predict that nulls should occur (within the range 0 to 60 degrees as follows:

$$\phi_P = 30 \text{ degrees}$$

$$\phi_G = 0,\ 11.5,\ 23.6,\ 36.9,\ 53.1 \text{ degrees}$$

Fig. 6.3L clearly shows these minima. Moreover, note from Fig. 6.3L that the upper envelope of gain falls off substantially with increasing elevation angle; this general result is caused by the natural free space directivity of the individual Yagi antennas. Note that this effect is much more pronounced for the larger six-element Yagi antennas (Fig. 6.3L) than for the smaller three-element equivalent stack (Fig. 6.2L).

Thus, the overall *H*-plane pattern is the result of three effects: first, the natural free space directivity of the individual Yagi antennas; second, the interference effect of the two real antennas; and third, the interference effect of the above ground system with its image counterpart. All three effects have different angular dependencies; it is therefore not surprising that the overall resultant can be quite varied and complex.

For those readers interested in constructing a vertically stacked Yagi antenna array, a careful scrutiny of Tables 6.1 and 6.2 and especially all of the relevant figures is quite enlightening. It is apparent that there is no single ideal design; nevertheless, there are a number of salient points that are worth noting.

1) A mast height (upper antenna height) of 0.75 wavelengths is really not high enough to get very much additional gain from stacking, especially with large Yagi antennas.
2) The higher systems provide better low-angle performance than the lower systems but sacrifice (sometimes needed) high-angle performance. They also provide less gain sacrifice due to ground images for big antennas and through increased antenna spacing provide less spoiling of the inherently good individual Yagi free space characteristics.
3) The important (lowest) first lobe gain is only weakly dependent on the placement of the lower antenna. The gain alone would favor h_L somewhat above $h_U/2$ (see for example Figs. 6.3G, 6.3H, and 6.3I); nevertheless, a lower placement (wider element spacing) will result in smaller beam interactions.
4) Mutual coupling or interaction between Yagi antennas tends to spoil the otherwise excellent properties of a single Yagi. This spoiling is most pronounced for low systems where spacings are small, not only to ground but between Yagi antennas (see for example Figs. 6.2A, 6.2B, and 6.2C). This spoiling can be easily seen in the altered patterns of the lower beam when the upper beam is physically present (dotted line) and when the upper beam is absent (broken solid line). You can also see the effect that stacking has on the *F/B* ratios (Tables 6.1 and 6.2) and also (not shown) the effects on the calculated driving-point impedances of both upper and lower Yagi antennas.
5) Interactive effects are also more serious when large Yagi antennas are used. This general result is anticipated and is due to the larger number of adjacent parasites; it is illustrated by comparing the dotted and broken-solid curves of Figs. 6.2B and 6.2C with those of Figs. 6.3B and 6.3C.
6) Any good stacked array (for example, Fig. 6.3G) will benefit by the Both, Lower, Upper (*BLU*) switch arrangement where at high angles a fill in the performance can be made (usually) using the lower antenna only. Best higher angle fill occurs when the placement of the lower antenna is at or preferably below $h_U/2$. A good practical height is $h_U/3 < h_L < h_U/2$. Note that a good fill obtained in this way slightly compromises maximum gain; however, this compromise is really not very serious.
7) With the *BLU* switch available it is interesting to compare perfor-

mances. In all cases, at the very lowest angles, Both and Upper give essentially identical results, that is, the stack is just as good as the upper antenna alone. However, the stack always accepts a broader range of vertical angles in its first lobe (due to its lower average height) and at its peak has more gain than either upper or lower alone. This gain advantage is one to three dB depending on the particular stack. Although this may not seem very impressive, experience demonstrates that the stack can indeed provide a commanding performance advantage over a single Yagi antenna and, coupled with the broader vertical coverage of the first lobe, will be more consistent.

8) A number of excellent stacked arrays can be chosen from these figures. As a good example note Fig. 6.3D. I have operated a stack very much like this on 14 MHz for several years; experience shows this to be a superb performer even without a *BLU* switch arrangement. Figures 6.3E and 6.3F also look very attractive, but the closer beam spacing results in increased variations in *F/B* properties and probably would require a *BLU* switch for best high-angle fill. For a higher stack note the excellent gain performances of Figs. 6.3G through 6.3I. However, for any of these cases, a fill seems desirable by the use of a *BLU* switch; note that for best fill at some higher angles the *upper* antenna should be used. For a very high stack Fig. 6.3J provides exceptional stacked gain, and by the additional use of the lower antenna for fill, it accommodates radiation angles up to nearly 30 degrees. However, at the 30 degree angles the system performance is abysmal, giving essentially zero response for any setting of the *BLU* switch.

Phase Derived Fill

I shall now turn briefly to an alternative method of obtaining higher-angle fill, a method that promises to be operationally simple and potentially very effective. Up to this point, I have used identical driver currents in both magnitude and phase. Let us now consider what effect is made on stacked *H*-plane patterns if the phase of the drive current in the lower antenna is changed relative to that in the upper antenna. I shall use as a test case the stack of Fig. 6.3D and will change the relative phase angle of the lower antenna drive current with respect to that of the upper antenna drive current. Computation of system performance under these conditions exhibits some remarkable effects. Table 6.3 shows performance as a function of the phase shift for several discrete relative phase angles from zero to 330 degrees, and Fig. 6.5 shows the *H*-plane patterns corresponding to five values of phase shift.

The *H*-plane pattern for any positive value of phase angle is nearly identical to the pattern for the same negative value; minor differences (which are also evident in Table 6.3) are caused by the detailed way in which all mutual coupling effects take place. It is easy to see from Fig. 6.5 that reversing the phase (shift = 180 degrees) results in excellent system performance at

Table 6.3 — Performance of the stack shown in Fig. 6.3D versus relative phase angle between lower and upper drive currents.

phase (deg)	*gain (dBi)*	*angle (deg)*	*F/B (dB)*	*resistance (ohms)*	*reactance (ohms)*
0	17.47	11	22.5	21.6	−0.2
30	17.27	11	23.0	21.6	−0.1
60	16.48	11	23.5	21.5	0.1
90	15.02	10	24.2	21.5	0.1
120	15.80	28	21.2	21.4	0.2
150	16.86	28	23.0	21.3	0.1
180	17.25	28	24.9	21.2	0.0
210	17.03	28	27.3	21.2	−0.1
240	16.16	28	31.0	21.2	−0.2
270	14.52	28	39.4	21.3	−0.2
300	16.15	11	21.3	21.4	−0.3
330	17.11	11	22.0	21.5	−0.2

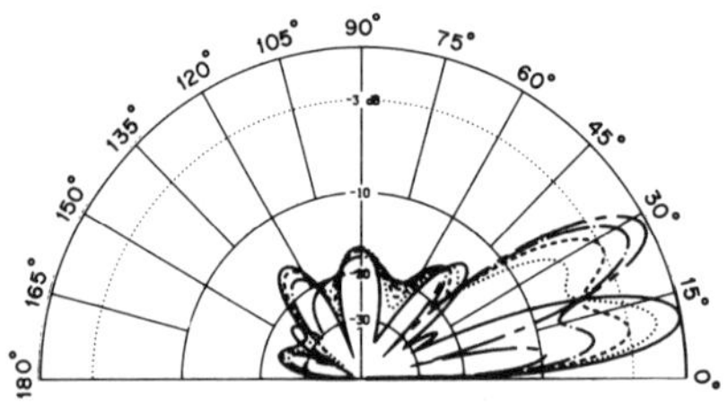

phase angle (deg)	**curve**
0	**solid**
60	**dots**
90	**dashes**
120	**chain-dots**
180	**chain-dashes**

Fig. 6.5 — *H*-plane radiation patterns for a two-array stack of six-element Yagis with out-of-phase excitation.

higher angles; basically giving maxima where the original *H*-plane pattern showed minima. At intermediate values of ϕ an intermediate result is obtained which is a combination of both the in-phase pattern (zero degrees phase angle) and the out-of-phase pattern (180 degree phase angle). Note that this higher angle fill effectively uses the extra gain potential of both Yagi antennas; it is therefore potentially superior to a single Yagi antenna fill and is also quite easy to implement (by switching in to only one of the antennas a coaxial line whose electrical length is one-half wavelength).

One can also see clearly from Fig. 6.5 that if the phase angle is relatively small, little degradation of system performance occurs; this fact potentially allows the stacking of dissimilar Yagi antennas. Nevertheless the use of dissimilar antennas raises questions about how to measure the effective phase shift and certainly increases the complications of controlling its value over a reasonable bandwidth of frequencies.

It is important to note that only two values of the phase angle are desired. The in-phase case (zero degrees) is best for low-angle performance and the out-of-phase case (180 degrees) is best for higher values of elevation angle.

All other values give inferior results to either one or the other of these cases.

Three and Four-Array Stacks

Let me now consider the possibility of stacking more than two Yagi antennas. It is obvious that some additional performance improvement should be possible provided the mast height is sufficiently high. As examples I show computations for two different evenly spaced stacks shown in Table 6.4. The first stack shows three six-element Yagis evenly spaced with a top height of 2.25 wavelengths, and the second shows four six-element Yagis evenly spaced to a top height of three wavelengths. Figs. 6.6 and 6.7 show the *H*-plane

Table 6.4 — Performance of three- and four-array stacks of six-element Yagis.

					all			*lowest*		
Fig.	h_1 (λ)	h_2 (λ)	h_3 (λ)	h_4 (λ)	*gain (dBi)*	*angle (deg)*	*F/B (dB)*	*gain (dBi)*	*angle (deg)*	*F/B (dB)*
6.6	0.75	1.50	2.25		19.20	7	14.9	15.27	20	15.9
6.7	0.75	1.50	2.25	3.00	20.24	6	16.0	14.93	19	12.4
					top			*lowest alone*		
Fig.	h_1 (λ)	h_2 (λ)	h_3 (λ)	h_4 (λ)	*gain (dBi)*	*angle (deg)*	*F/B (dB)*	*gain (dBi)*	*angle (deg)*	*F/B (dB)*
6.6	0.75	1.50	2.25		16.15	5	22.3	15.60	18	21.5
6.7	0.75	1.50	2.25	3.00	16.11	5	21.6	15.60	18	21.5

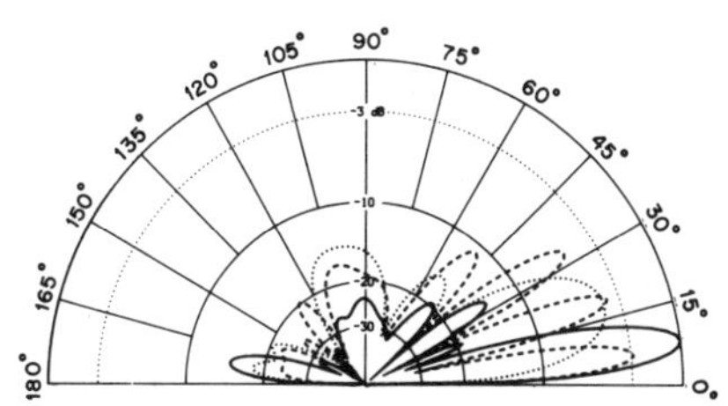

combination	curve
all	solid
lowest	dots
highest	dashes

Fig. 6.6 — *H*-plane radiation patterns for a three-array stack of six-element beams at 0.75, 1.5 and 2.25 wavelengths above ground.

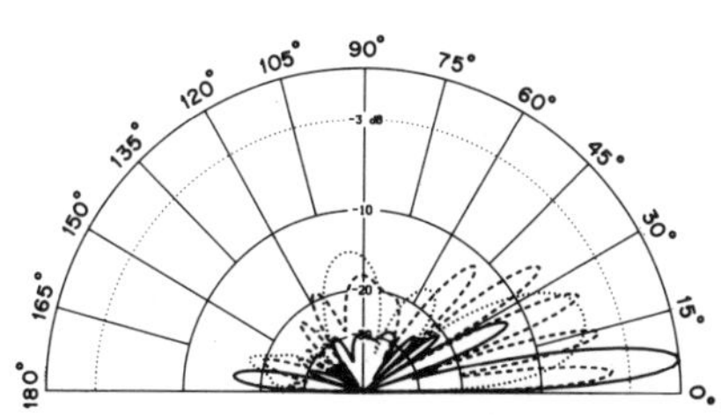

combination	curve
all	solid
lowest	dots
highest	dashes

Fig. 6.7 — *H*-plane radiation patterns for a four-array stack of six-element beams at 0.75, 1.5, 2.25 and 3 wavelengths above ground.

patterns for these two systems; they should be compared with Figs. 6.3G and 6.3J, which are basically equivalent two-Yagi stacks of comparable mast height. It is at once apparent from Fig. 6.6 that the addition of the third Yagi antenna gives a main lobe gain over Fig. 6.3G of 1.2 dB; all other characteristics are quite comparable. Likewise, Fig. 6.7 shows that the two additional Yagi antennas give a main lobe gain increase of 2.9 dB over Fig. 6.3J; again, all other characteristics are quite similar.

These examples of vertically stacked Yagi antenna arrays using more than two antennas show that a noticeable gain increase is possible over a two-antenna stack; moreover they open up a wide range of higher-angle fill possibilities. As an example, Fig. 6.7 shows patterns where all four antennas are excited; for fill at higher angles, the lowest antenna (dotted line) and the highest antenna (dashed line) might be used alone. Note, however, that additional fill situations are possible if two or even three of the original four antennas are excited coherently. Moreover one can also consider feedline phasing for even better fill. Clearly, a host of possibilities exists, but the practical use of all potentially desirable combinations not only requires a complex switching system but a great deal of trouble in determining experimentally the right combination for the prevailing circuit conditions. Surely the additional complexity and expense of these large vertical stacks reaches a point of practical diminishing returns. Nevertheless, how fortunate we are to be able to predict with reasonable confidence the performance of such large systems, without ever having to build one.

Optimization of Stack Arrays

I shall conclude this chapter on stacking by referring again to the basic two-antenna stack shown in Fig. 6.3D. Note that the *F/B* performance has deteriorated from the excellent free space performance of the individual Yagi antennas of 47 dB to 22.7 dB. Analogous to the optimization of a single Yagi antenna over ground, it is possible to optimize the basic Yagi design for this stacked system. Table 6.5 shows the optimized parameters of the six-element

Table 6.5 — Element lengths and positions for two-array stacks of six-element beams, optimized in free space and for the heights shown in Fig. 6.3D, by shifts in boom positions for the first and third directors. All element radii are 0.00052599 wavelengths.

	free space		*heights = 0.6 λ and 1.5 λ*	
element	*length (λ)*	*position (λ)*	*length (λ)*	*position (λ)*
reflector	0.49483	0.000	0.49483	0.000
driver	0.48073	0.150	0.48123	0.150
D1	0.44771	0.29832	0.44771	0.30184
D2	0.44771	0.450	0.44771	0.450
D3	0.44771	0.60150	0.44771	0.64120
D4	0.44771	0.750	0.44771	0.750

Table 6.6 — Performance of the free space optimized two-array stack of six-element Yagis described in Table 6.5.

normalized frequency	*gain (dBi)*	*F/B (dB)*	*angle (deg)*	*resistance (ohms)*	*reactance (ohms)*
0.996	10.58	26.6	0	22.0	−6.2
0.998	10.64	32.7	0	21.6	−3.1
1.000	10.70	134.8	0	21.3	−0.1
1.002	10.75	32.7	0	21.0	3.1
1.004	10.80	26.7	0	20.8	6.3

Table 6.7 — Performance over ground of the optimized two-array stack of six-element Yagis described in Table 6.5.

normalized frequency	*gain (dBi)*	*F/B (dB)*	*angle (deg)*	*resistance (ohms)*	*reactance (ohms)*
0.996	17.26	24.8	11	16.3	−6.5
0.998	17.34	30.8	11	15.8	−3.4
1.000	17.40	93.0	11	15.3	−0.2
1.002	17.47	30.6	11	14.8	3.0
1.004	17.52	24.5	11	14.4	6.2

Yagi antenna first for free space and second for the stack of Fig. 6.3D (h_L = 0.6 wavelengths, h_U = 1.5 wavelengths). Tables 6.6 and 6.7 show the swept-frequency performance of each of these cases close to the design frequency.

The iterative optimization was carried out by adjustments of the boom positions of the first and third directors to obtain high *F/B* (more than 90 dB), and by a slight adjustment of driven-element length to minimize reactance at the design frequency. Note again that, because of mutual coupling interactions, the stacked Yagi antenna is not the same Yagi antenna as it would be in free space, nor is it the same Yagi antenna as it would be singly over ground (compare with Table 5.4).

Orthogonal and Antiparallel Stacked Yagis

Now that an optimized Yagi design has been found, which provides the superlative performance shown in Table 6.7 when the two stacked antennas are parallel to each other, we can ask about performance degradation when the two Yagi antennas are orthogonal to each other and also when they are antiparallel. This question is relevant when a stack is used where the lower antenna is fixed in some direction and the upper antenna is rotatable. Table 6.8 shows the system performance for the case where both Yagis are supported at the center of the boom.

It is clear that, in principle, optimization can be carried out for only one configuration, and performance will automatically deteriorate somewhat for other geometries. The extent of deterioration will be more severe for stacks with small antenna spacings (both with respect to ground and to each other); that

Table 6.8 — Performance of a two-array stack of six-element Yagis for different relative boom orientations.

	Both (angle = 11°)				*lower (angle = 21°)*			
config-uration	*gain (dBi)*	*F/B (dB)*	*R (ohms)*	*X (ohms)*	*gain (dBi)*	*F/B (dB)*	*R (ohms)*	*X (ohms)*
parallel	17.40	93.0	15.3	−0.2	15.07	31.2	13.3	−0.6
orthogonal					15.14	36.9	13.7	−1.0
antiparallel					15.00	27.1	13.3	−0.6

	upper (angle = 9°)			
config-uration	*gain (dBi)*	*F/B (dB)*	*R (ohms)*	*X (ohms)*
parallel	16.34	25.5	15.2	−0.8
orthogonal	16.59	20.8	15.4	−1.1
antiparallel	16.39	24.0	21.0	1.9

is, with lower overall mast height. For the stack shown in Table 6.8, it is gratifying to see that the performance for all situations is really quite acceptable.

Summary

1) Vertical stacking of two Yagi antennas allows both substantial improvement in low-angle system performance and improved flexibility. This flexibility can be used either to obtain fill at some needed higher angles or to illuminate other azimuthal angles (one or two Yagi antennas rotatable).

2) Mast heights of between one and perhaps 2.5 wavelengths can provide excellent two-Yagi stacked systems.

3) Higher masts favor low-angle radiation and also give smaller mutual interaction effects. However, they also treat the (occasionally useful) higher angles unfavorably.

4) For all vertical stacks, improved performance is available if excitation is switchable to both antennas, the lower antenna, or the upper antenna (a "*BLU*" switch). Switching must be done in a way that preserves phase integrity and keeps the total drive impedance matched to the supply coaxial line. For those antenna stacks where the lower and upper beams remain aligned (rotate together), a highly useful switch is a phase reverser to only one of the beams.

5) Vertical stacks using three or four Yagi antennas can display even greater performance, but the stacks must be very high and must use for best results a more complex feedline switching arrangement.

6) Optimization (very high *F/B* at one frequency) can be obtained for only one physical configuration at a time. Nevertheless, there are practical examples where an optimized antenna design for a two-Yagi stack will still exhibit excellent properties when only the lower antenna or only the upper antenna is excited; moreover, these excellent properties are retained even if the azimuthal directions of the two individual Yagi antennas are parallel, orthogonal or even antiparallel.

CHAPTER 7

PRACTICAL DESIGN

Up to this point the specifications for a Yagi antenna have been made only in terms of strictly cylindrical elements. Each element is characterized by a position along the boom, a physical length, and a radius; all three of these quantities are expressed in wavelengths at a central design frequency. Such specifications have led to a number of rather good antenna designs and I shall shortly list a brief selection of such designs.

However, when a real Yagi antenna is constructed it will rarely be convenient to rigorously adhere to the cylindrical element design. To start, the element diameter is usually adjusted to fit a mechanical requirement (wind loading, etc.); moreover, the element itself is usually not a cylinder but a series of telescoping tubes starting with a large diameter section at the boom and "tapering" to a small diameter section at the outer end of the element. In addition, the element is fastened to the boom with a boom clamp which may be a plate or angle bracket U-bolted to both boom and element. Some mechanical designs even put the element directly through the boom.

Thus, the path from the cylindrical design to a practical antenna will involve three tasks: first scaling the original design to an equivalent new design using a different (average) element radius; second, computing the change in element length to account for the effect of tapered element section radii; and third, making (usually minor) corrections to allow for the effect of the element-to-boom clamping system. Methods for carrying out each of these three tasks will be given below, following a discussion of preferred antenna designs.

Preferred Antenna Designs

I shall show in this section one preferred design example for each case of two, three, four, five and six-element Yagi antennas. Recall from Chapter 2 that simplistic Yagis (element spacing uniform and all directors having a common length) are as good as any other design up to a boom length of one wavelength. It was shown that a good two-element beam would have a boom length of about 0.15 wavelengths; the exact length is not critical and is a compromise between better gain and lower efficiency and bandwidth. Best parasite element length is a compromise between better forward gain and lower *F/B* ratio. For a three-element beam it was shown that a boom length of about 0.3 wavelengths produced a naturally high *F/B* and similarly for four, five, and six-element beams a boom about 0.75 wavelengths gives a naturally good *F/B* ratio. Table 7.1 shows the characteristics of these good Yagi designs. It should be emphasized that these particular antenna designs are not unique; for example, the boom length can be varied somewhat (see Chapter 2). Longer booms generally give larger forward gain, but make the

Table 7.1 — Characteristics for preferred Yagi designs. All dimensions are given in wavelengths at the design frequency, and all element radii are 0.00052599 wavelengths.

	two-element		*three-element*		*four element*	
element	*position* (λ)	*length* (λ)	*position* (λ)	*length* (λ)	*position* (λ)	*length* (λ)
reflector	0.0	0.49366	0.0	0.49404	0.0	0.49136
driver	0.150	0.47050	0.150	0.48572	0.250	0.47895
D1			0.300	0.46525	0.500	0.46273
D2					0.750	0.46273
D3						
D4						
Gain (dBi)	6.9		8.3		10.6	
F/B (dB)	8.0		31.2		44.0	

	five-element		*six-element*	
element	*position* (λ)	*length* (λ)	*position* (λ)	*length* (λ)
reflector	0.0	0.49944	0.0	0.49478
driver	0.1875	0.48031	0.150	0.48038
D1	0.375	0.45187	0.300	0.44766
D2	0.5625	0.45187	0.450	0.44766
D3	0.750	0.45187	0.600	0.44766
D4			0.750	0.44766
Gain (dBi)	10.5		10.7	
F/B (dB)	32.4		46.8	

Table 7.2 — Element lengths and positions for six-element Yagis optimized at different heights. All dimensions are given in wavelengths at the design frequency, and all element radii are 0.00052599 wavelengths.

	free space		*1.0λ*		*0.6λ and 0.5λ*	
element	*position* (λ)	*length* (λ)	*position* (λ)	*length* (λ)	*position* (λ)	*length* (λ)
reflector	0.0	0.49478	0.0	0.49478	0.0	0.49478
driver	0.150	0.48038	0.150	0.48000	0.150	0.48140
D1	0.29884	0.44766	0.3074	0.44766	0.30215	0.44766
D2	0.450	0.44766	0.450	0.44766	0.450	0.44766
D3	0.60189	0.44766	0.5980	0.44766	0.64165	0.44766
D4	0.750	0.44766	0.750	0.44766	0.750	0.44766

frequency for highest *F/B* ratio drop somewhat below the center of the band where gain remains high.

It has also been shown (see Chapter 3) that there exists a procedure which allows fine tuning the *F/B* ratio; this optimization procedure can be done for Yagi antennas having four or more elements. Optimization must be carried out for a specific end use. Table 7.2 shows optimized six-element beams first for free space use, next for operation at one wavelength above ground and finally for operation in a two-Yagi stack at heights of 0.6 and 1.5 wavelengths. These parameters are mathematically correct, but note from Chapter 1 that approximations used in the model really do not justify complete confidence in the precise values in Table 7.2. Nevertheless I suspect that practical antennas (for use over ground) constructed from this table will exhibit superior properties to the (free space) six-element case shown in Table 7.1.

Radius Scaling

Any Yagi antenna design, such as those shown in Table 7.1, can be scaled either to other center frequencies or to elements of different diameter at the same center frequency. Since all design parameters are dimensions expressed in wavelengths at a central design frequency the values shown in Table 7.2 are invariant to frequency scaling and therefore the behavior of the antenna will be unaffected by the choice of central design frequency. However this is true only if *all* physical dimensions (including element radius) are adjusted in proportion to the desired wavelength.

Experience shows that practical element radii expressed in wavelengths are not constant; at low frequencies (long wavelengths) relatively thin elements are used while at high frequencies relatively fat elements are normal. How then can a given design be altered to an equivalent design where element radius is changed? The clue is to make the impedance of the changed element exactly the same as the impedance of the original element; in this way exactly the same element currents will flow, resulting in the same detailed antenna performance.

Since the (radiation) resistance of the element is essentially unaffected by changes in radius, we need only make the reactance invariant to scaling element radius. Recall from Chapter 1 that element reactance at the normalized central design frequency, $F = 1$, can be written:

$$X = A\,(1 - F_R) \tag{7.1}$$

where F_R is the normalized resonant frequency and

$$A = 430.8 \log K - 339 \tag{7.2}$$

Here K is the ratio of wavelength (at the design frequency) to element radius: λ/a. The resonant frequency depends on the resonant length, l_R, through $F_R = l_R/l$, where l is element length. The resonant length depends only on radius (through K) and can be written:

$$l_R \approx 0.5 - [33.25 + 3.19 \log K - 0.35(\log K)^2]/[861.6 \log K - 678] \tag{7.3}$$

If we wish to scale the element radius from its original value to a new value we must ensure that X is unchanged. Denote the original ("design") element by subscript d and the new ("scaled") element by a subscript s. We are given l_d, a_d, and a_s, and we wish to find the new length l_s such that:

$$X_s = X_d \tag{7.4}$$

The solution is as follows. First, compute A_d and A_s from (7.2), and l_{R_d} and l_{R_s}, from (7.3). Then solve (7.4) for F_{R_s}, in terms of known quantities using (7.1):

$$F_{R_s} = 1 - A_d(1 - F_{R_d})/A_s \tag{7.5}$$

Finally, compute l_s from the resonant length and resonant frequency for the scaled element:

$$l_s = l_{R_s}/F_{R_s} \tag{7.6}$$

These three steps give the length of the scaled element (with new radius) which has the same reactance as the original element. It is quite simple and most convenient to set up the entire conversion procedure represented by (7.2) through (7.6) on a small programmable calculator.

An example will illustrate the nature of results. Consider the antenna design for the six-element antenna of Table 7.1; this would be a reasonable design for a 14.2 MHz antenna where one wavelength is 831.76 inches and a radius of 0.00052599 wavelengths corresponds to an element diameter of 7/8 inches. This would be a reasonable dimension for a mechanically adequate element. Now suppose that we would like an equivalent antenna for 28 MHz where the radius should be increased. Table 7.3 shows the computations. The original design is shown as case 1, the scaled design for an intermediate radius of 0.0008 wavelengths (about 5/8 inches diameter) as case 2, and a one inch diameter (0.0012 wavelengths radius) element as case 3. Note that the changed values for element lengths are not wholly intuitive because two things happen simultaneously. As the radius increases the Q drops, requiring a greater spread in the resonant fre-

Table 7.3 — Element lengths and resonant lengths in wavelengths, normalized resonant frequencies, and reactances in ohms, for three values of element radius: case 1 = 0.00052599λ, case 2 = 0.0008λ, case 3 = 0.0012λ.

	1	*2*	*3*
		reflector	
l	0.49478	0.49445	0.49408
l_R	0.48140	0.48002	0.47847
F_R	0.97295	0.97082	0.96841
X	29.0	29.0	29.0
		driver	
	1	*2*	*3*
l	0.48038	0.47892	0.47729
l_R	0.48140	0.48002	0.47847
F_R	1.00212	1.00230	1.00247
X	−2.3	−2.3	−2.3
		director	
	1	*2*	*3*
l	0.44766	0.44393	0.43977
l_R	0.48140	0.48002	0.47847
F_R	1.07537	1.08130	1.08801
X	−80.9	−80.9	−80.9

quencies of reflector and director. However, at the same time the resonant physical length also changes.

It is important to reiterate that it is conceptually *wrong* to scale boom length and element lengths (for example, to convert a VHF antenna design to HF) without also scaling the element radius. The correct way to adjust an antenna element when the old and new radii (in wavelengths) are different is to modify *both* length and radius to give the same electrical reactance.

We now have the tools to convert a given antenna design such as shown in Table 7.1 to a new antenna design where the element radius is changed; the new antenna will perform exactly the same as the original antenna at the central design frequency. However the frequency-swept behavior of the new antenna, while qualitatively similar to the original, will show a broader or narrower bandwidth depending on the change in element Q.

Taper Corrections

Up to this point all antenna calculations have been made for strictly cylindrical elements and the results will apply directly to most high-frequency (small) Yagi antennas where the general practice is to use cylindrical elements. However, for frequencies less than about 30 MHz, mechanical considerations usually require that the element consists of one or more telescoping sections of tubing. At the lower frequencies (say 7 MHz and below) the Yagi antenna becomes gigantic and it is no small mechanical engineering task to even construct a good element. Small diameters favor smaller wind forces but are insufficiently rugged for long elements. It is therefore a universal practice that these large elements be made of several telescoping sections; the largest diameter section is clamped to the boom and succeeding monotonically smaller diameter sections make up the outer portions of the element.

Thus, it is important to understand how to relate the actual detailed taper schedule of an element (that is, the diameters and lengths of all sections) to the *equivalent length* of a cylindrical element having the same average diameter. "Equivalent" means that the resonant frequency and the Q are the same for the tapered element and the equivalent cylinder.

To start, I shall introduce the concepts of pipe inductance and pipe capacitance. Consider a cylindrical element of total length s and radius a as shown in Fig. 7.1. Define a position coordinate, x, whose origin is at the center of the element, and the related angle coordinate, θ, where $\theta = \pi x/s$ in radians. Excitation of this element in the neighborhood of the resonant frequency will produce sinusoidal current and voltage distributions:

$$I(\theta) = I_0 \sin(2\pi f t)\cos\theta \tag{7.7}$$

and

$$V(\theta) = V_0 \cos(2\pi f t)\sin\theta \tag{7.8}$$

The driving-point impedance of the element consists of a resistance (which is directly related to far-field energy radiation) and of course a reactance. All reactance effects including resonant frequency and electrical Q are caused by near-field (non-radiation) energy storage. Energy storage occurs in two ways: the magnetic flux surrounding the current distribution in (7.7) and the electrical field produced by the voltage distribution in (7.8). Note that at certain instantaneous times ($t = n/2f$) the current everywhere is zero and all stored energy resides in the electric field. Similarly at certain other times ($t = n/2f + 1/4f$) the electric field vanishes and all of the stored energy resides in the magnetic flux. As time progresses the (constant) total stored energy transfers back and forth between magnetic and electrostatic fields. The transfer frequency is of course the resonant frequency of the element. Note that as a result of this complete non-radiative energy transfer the peak magnetic stored energy must exactly equal the peak electrostatic stored energy. Note also that the resonant exchange frequency must decrease as the total stored energy is increased.

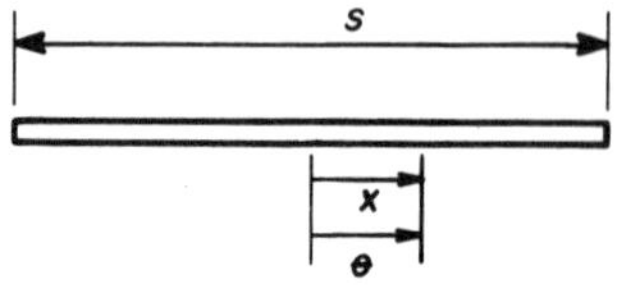

Fig. 7.1 — Coordinates of a single Yagi element.

Now consider the effect of inserting an infinitesimal length of pipe of the same radius into the element of Fig. 7.1, at the center. Recall from Chapter 1 that the original element driving-point reactance is:

$$X = (430.8\log K - 339)(F - F_R) \tag{7.9}$$

At the resonant frequency the reactance vanishes. Inserting an additional infinitesimal length of pipe, Δs, at the center will change the resonant frequency to F^*_R. At this new frequency the total reactance again vanishes. The added reactance due to the inserted pipe must be balanced by the original pipe reactance at the new frequency:

$$0 = (430.8\log K - 339)(F^*_R - F_R) + 2\pi f \Delta L \tag{7.10}$$

where ΔL is the increased inductance due to Δs.

Note that the pipe inserted at the element center can produce only inductive effects (stored magnetic flux) since the electrial potential is strictly zero. Now F^*_R is clearly related to F_R by the overall length of the element:

$$F^*_R - F_R = -\Delta s/s \tag{7.11}$$

from which

$$\Delta L/\Delta s = (430.8\log K - 339)/(2\pi f s)$$

or, since $s \approx \lambda/2$,

$$\Delta L/\Delta s = (430.8\log K - 339)/(\pi c) \qquad (7.12)$$

where c is the velocity of light. Thus the addition of the small infinitesimal pipe section causes the element to behave just as though a pure series inductance were added. The effective inductance per unit length, which I designate by *IND*, is given by (7.12), and is easily expressed in henrys/meters:

$$IND = (43.08\log K - 33.9)(1.061 \times 10^{-8}) \qquad (7.13)$$

Recall now that in a simple model of a resonant circuit it is easy to relate the magnitude of voltage on the reactive components to the magnitude of input current:

$$|V_0| = |I_0| RQ \qquad (7.14)$$

and also recall from Chapter 1:

$$2RQ = (430.8\log K - 339) \qquad (7.15)$$

Now consider extending the element in Fig. 7.1 by a pipe of length Δs of the same radius at its outer end ($x = s/2$). Here the current is zero so that the small pipe can act to increase the electrostatic energy only (capacitive effect). Since in this case (7.9) is still valid, the total increase in stored energy should be just the same as it was for insertion at the center. Therefore:

$$\Delta L(I_0)^2/2 = \Delta C(V_0)^2/2 \qquad (7.16)$$

where ΔC is the capacitance increase due to Δs at the element end and ΔL is the increase in inductance due to Δs at the element center. From (7.14) and (7.16):

$$\Delta C = \Delta L/(RQ)^2 \qquad (7.17)$$

Using (7.15) (7.17) and (7.12):

$$\Delta s/\Delta C = (43.08\log K - 33.9)(25\pi c/10) \qquad (7.18)$$

or in meters/farad:

$$\Delta s/\Delta C \equiv 1/CAP = (43.08\log K - 33.9)(2.356 \times 10^9) \qquad (7.19)$$

where *CAP* is the capacitance per unit length. Note that $1/CAP$ is directly related to *IND*, differing only in a constant multiplier.

Thus we now can think of a cylindrical section of element pipe as contributing to element inductance (7.13) and element capacitance (7.19). Each contribution is a function of K (and therefore radius), and each will depend on the current or voltage on the pipe section.

Let us now see what happens if a small section of pipe length $\Delta B/2$ is first removed at a position x (or corresponding θ) and for symmetry also at $-x$

from the element shown in Fig. 7.1. Now replace these removed sections with equal length sections of larger radius. Thus the overall length of the element remains at s but cylindrical "bumps" occur at x and $-x$. As a result of these bumps the stored energy of the system is changed and therefore the resonant frequency is changed. The bumps' contribution to stored energy, W^*, will be:

$$2W^* = \Delta B[IND^*(I_0^2\cos^2\theta) + CAP^*(V_0^2\sin^2\theta)] \tag{7.20}$$

The relationship of V at the end of the element to I at the center is essentially unchanged from the original element (see (7.16)):

$$CAP\ V_0^2 = IND\ I_0^2 \tag{7.21}$$

Note also from (7.13) and (7.19) that

$$CAP^*/CAP = IND/IND^* \tag{7.22}$$

so that (7.20) can be rewritten:

$$2W^* = \Delta B[IND^*(I_0^2\cos^2\theta + (IND^2/IND^*)(I_0^2\sin^2\theta)] \tag{7.23}$$

Let us now find an equivalent length, $\Delta A/2$, of the original pipe which, placed at the same positions as each of the bumps, contributes an equal stored energy.

$$2W = \Delta A[IND\ I_0^2(\cos^2\theta + \sin^2\theta)] = \Delta A\ IND\ I_0^2 = 2W^* \tag{7.24}$$

so that

$$\Delta A/\Delta B = (IND^*/IND)\cos^2\theta + (IND/IND^*)\sin^2\theta \tag{7.25}$$

Now for a longer section (longer bump) going from θ_1 to θ_2 the equivalent length of the original pipe size can be easily calculated. Designate the ratio IND^*/IND as m, the length of the long bump as s_B, and the length of the original pipe which gives the same stored energy as s_A. Then:

$$s_A/s_B = m\ \overline{\cos^2\theta} + \frac{1}{m}\ \overline{\sin^2\theta} \tag{7.26}$$

The angular functions are to be averaged over the complete bump section; the result is:

$$s_A/s_B = (m + \frac{1}{m})/2 + (m - \frac{1}{m})f(\theta)/2 \tag{7.27}$$

where

$$f(\theta) = (\sin 2\theta_2 - \sin 2\theta_1)/(2\theta_2 - 2\theta_1) \tag{7.28}$$

and θ is measured in radians.

Thus we can now compute, from a given element taper schedule involving several sections with different pipe diameters, the equivalent lengths of sections of "standard" cylindrical pipe. The procedure is to choose first the standard cylinder radius which is to provide equivalent Q. A good choice is the pipe size at the center of each half-element. Next, for each section of the tapered element compute the starting θ_1, and ending θ_2. Then, for each section compute m; it is easily derived from (7.13), i.e.:

$$m = (43.08\log K^* - 33.9)/(43.08\log K - 33.9) \qquad (7.29)$$

From (7.29) and (7.27) compute s_A/s_B which, multiplied by the tapered section physical length, gives the equivalent section length of the standard pipe. Adding the lengths of all equivalent sections gives the overall length of the standard cylindrical element which should perform essentially the same as the chosen taper schedule.

Perhaps an example will illustrate the procedure. Fig. 7.2 shows schematically a half-element with five different sections whose physical diameter ranges from 1.25 inches at the boom to 0.5 inches at the outer end. Readers will recognize this taper schedule as one in common use (by Wilson) for a 14 MHz Yagi reflector antenna element. The middle pipe section (7/8 inches in diameter) will represent the standard pipe. At a frequency of 14.2 MHz one wavelength is 831.76 inches and thus $a = 0.00052599$ and $K = 1901.17$. Table 7.4 illustrates the way in which one calculates equivalent cylinder section lengths. For each section, column 2 shows the actual physical length, s_B, column 3 shows pipe diameter, column 4 the K^* value, column 5 the value of m computed from (7.29), column 6 values for θ_2, column 7 values of $f(\theta)$ computed from (7.28), and column 8 equivalent section lengths, s_A, com-

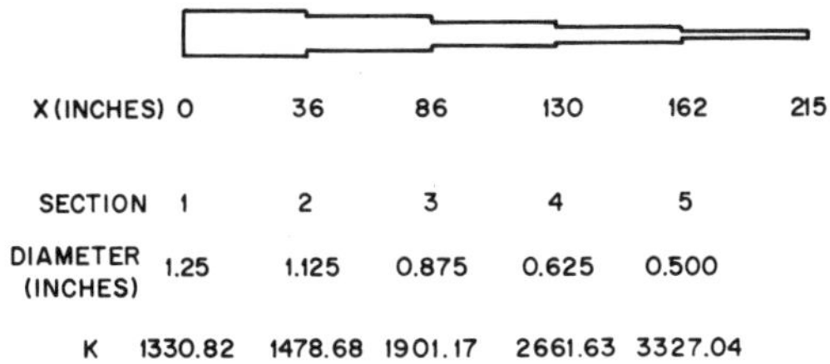

Fig. 7.2 — Example of a radius-tapered half-element.

Table 7.4 — Equivalent length computations for the element of Fig. 7.2.

section	s_B *(inches)*	$2a^*$ *(inches)*	K^*	m	θ_2 *(degrees)*	$f(\theta)$	s_A *(inches)*
1	36	1.250	1330.82	0.93784	15.070	0.95452	33.868
2	50	1.125	1478.68	0.95620	36.000	0.61449	48.674
3	44	0.875	1901.17	1.00000	54.419	−0.00718	44.000
4	32	0.625	2661.63	1.05864	67.814	−0.52851	31.088
5	53	0.500	3327.04	1.09752	90.000	−0.90300	48.770
	215						206.400

puted by (7.27) and the column 2 values. The value for θ_1 for each section is simply the value of θ_2 for the preceeding section, with $\theta_1 = 0$ for the first section.

Note that the total length of the tapered half-element is 215 inches whereas the total length of the equivalent cylindrical half-element is only 206.40 inches. In other words, just due to the taper effect alone the total (full length) tapered element must be made 17.2 inches longer than an equivalent cylinder. This taper correction is surprisingly large; it shows clearly that length alone is a totally inadequate specification for an HF Yagi element. The physical reason why the tapered element must be longer than an equivalent cylinder is that the inner (larger) sections have smaller inductance than a standard cylinder and therefore must be made longer; similarly the outer (smaller) sections have smaller capacitance than the standard cylinder and must also be made longer. Therefore the taper correction will be quite small if the taper is small, but quite significant if the taper is large.

In the derivation of taper correction calculations, I have assumed that radial "bumps" are treated as small perturbations on the strictly cylindrical case and that the current and voltage distributions are sinusoidal. Note that m values for the heavily tapered element of Fig. 7.2 differ from unity by only a few percent; thus the calculation, even though made by a perturbation method, should be reasonably good. Moreover the current distribution should be reasonably sinusoidal over the tapered element. Nevertheless there may be some small inaccuracies in the overall calculation. It is important to note, however, that we are after a length correction of only a few percent and therefore some inaccuracy in the computation of the (small) correction is tolerable.

One further point merits elaboration. The procedure just outlined only allows a computation of cylinder equivalents from a given taper schedule; how may we compute a suitable taper schedule starting from a given cylinder? One good procedure is to initially specify all of the taper schedules from mechanical considerations, leaving as a variable only the length of the outermost section. Choose a guessed or estimated length for this section and compute the overall equivalent cylinder. It will generally miss the desired length by a differential length Δs. One can now readjust the length of the outermost section by $-m\Delta s$ and recalculate. One or two such iterations will bring the tapered element equivalent cylinder length into adequate agreement with the desired figure.

A second procedure is almost as accurate and far easier to apply. Assume that all elements can be constructed from the same basic taper schedule; variations are made by varying the length of only the outermost section. One can relate the tapered half-element length *for this schedule only* to equivalent cylinder half-length through:

$$l = pl_{eq} + q \tag{7.30}$$

To find values for the coefficients p and q, calculate equivalent lengths from two different tapered half-lengths, say for a reflector and a director, using the procedure demonstrated above. Then p and q are determined as:

$$p = (l_1 - l_2)/(l_{eq_1} - l_{eq_2}) \tag{7.31}$$

and

$$q = (l_2 l_{eq_1} - l_1 l_{eq_2})/(l_{eq_1} - l_{eq_2}) \tag{7.32}$$

Note that (7.30) allows a direct computation of taper half-length starting from a given desired cylinder length. These empirical equations are remarkably accurate and are much easier to use than the detailed iterative procedure.

Boom and Element Clamping Correction

I now come to the subject of the boom-to-element mechanical clamping system and its effect on the element reactance and hence resonance. It is clear that a wide range of clamping systems are in common use; it is virtually impossible to make valid calculations for all varieties. Nevertheless there are two major kinds and it is helpful to understand them.

The first clamping system is simply to put the element directly through the (round) boom. In this construction a length of element equal to the complete boom diameter is replaced with the boom itself. Since this replacement occurs at a current node we must determine the effective inductance of the replacement; once this is done it can be considered as the first section of a tapered element from which an equivalent cylinder length can be calculated. I have not attempted a rigorous calculation of boom inductance; instead I refer to the measurements of Viezbicke in which his Fig. 10 shows that element overall length due to the presence of a round boom should be increased by about 0.7 times the diameter of the boom.[1] This is tantamount to saying that the inductance of the boom section of the element is very low compared with normal element inductance; physically this is an expected result. The low inductance of course is caused by the boom blocking the magnetic flux which accompanies element current flow.

The second clamping system is much more widely used since it permits easier element maintenance and replacement. In this system either a flat metal rectangular plate or an angle bracket is interposed between element and boom; two U-bolts fasten the boom to the bracket and two more U-bolts fasten the element to the bracket. The U-bolts may also use saddles or cradles which are mechanically better and which further tend to separate boom and element. For this clamping system we wish to know the inductive effect of the boom itself, and more importantly the inductive effect of the bracket. I have found

[1]This and other footnotes are given at the end of this chapter.

experimentally that for this clamping system the boom itself has remarkably little effect. Even though the boom and element are in physical contact, the element length should be increased by only six percent of the boom diameter; this small correction rapidly disappears as the element is spaced away from the boom (even by a small amount). The reason this result is so different from the through-the-boom result is the relative ease with which magnetic flux, which results from element current flow, can squeeze between boom and element, especially if there is any gap between them.

The correction in length due to the mounting bracket is readily calculable. The method is to first calculate the equivalent radius of the element plus bracket (which produces the same inductance) and second to use this equivalent radius as the first short section of a taper schedule. This theory is given by Mushiake and Uda.[2] The equivalent radius, a_e, of a flat thin plate of total width, A, is simply:

$$a_e = A/4 \tag{7.33}$$

and that for a right-angled bracket of width A and height B is given by a rather complicated expression which depends only slowly on the ratio B/A. For ratios between 0.3 and 1.0 a good approximation (error less than five percent) is:

$$a_e \approx 0.2(A + B) \tag{7.34}$$

Mushiake and Uda also show how to calculate an equivalent radius for two parallel conductors. If S_1 and S_2 are the lengths of the peripheries of the cross sections, a_1 and a_2 are the equivalent radii of the two conductors, and d is the mean distance between them, the equivalent radius of the combination of both conductors is determined from:

$$\ln a_e = (S_1^2 \ln a_1 + S_2^2 \ln a_2 + 2S_1S_2 \ln d)/(S_1 + S_2)^2 \tag{7.35}$$

Equations (7.33) and (7.35) permit a calculation of the equivalent radius of an element which is proximate to a plate; similarly (7.34) and (7.35) provide a way of calculating the equivalent radius of an element proximate to an angle bracket. To check this method of calculation, I have determined the experimental detuning effect of a plate just touching a one inch diameter element resonant at 46 MHz. Table 7.5 shows both theoretical and experimental results for two different plates. These

Table 7.5 — Increase in resonant frequency due to a proximate plate. Element diameter is one inch and length produces resonance at 46 MHz.

	dimensions		*change in F_R*	
plate	*length (inches)*	*width (inches)*	*theory (percent)*	*experiment (percent)*
1	4.5	3.625	0.304	0.325 ± 0.1
2	6.0	4.0	0.530	0.521 ± 0.1

experiments were not particularly accurate because the exact resonant frequency is difficult to determine; nevertheless the agreement of theory and experiment within estimated experimental accuracy is gratifying.

Note that element length corrections due to a proximate mounting plate or bracket can easily be as much as 10 percent of the plate length. These corrections are not especially large in practice but should be made wherever there is a relatively large boom-to-element clamping system.

Examples of Three-Element Beams

It may be helpful to show explicitly how to specify a good three-element beam starting with the design data specified in Table 7.1. I shall first perform radius scaling, then taper schedule calculations, and finally apply reasonable boom clamping and boom corrections. To illustrate the full range of these computations I shall use this procedure to specify a 14.2 MHz beam, a 21.3 MHz beam, and finally a 28.5 MHz beam.

1) Choose an average element cylinder size which is sufficiently strong. I shall assume that the final element is made of good strong seamless aluminum tubing such as 6061-T6 with 0.058 inch wall thickness. For all three bands I choose a cylinder size of 7/8 inches outside diameter, although for 28.5 MHz a somewhat smaller size is probably permissible.

2) Choose a convenient taper schedule which is easily made from standard 12 foot lengths, leaving the length of the outermost section to be adjusted for correct overall length. The sections of seamless tubing (excepting the last section) are slit back about three inches at the outer ends (I use one slit only) and a common stainless steel hose clamp fastens sections together. Tubing overlap of about eight inches gives good joint strength. For 14.2 MHz the second section is actually a full 12 foot section over which is slid the shorter first section; this procedure gives added strength at the center and improves the ease of clamping with U-bolts and saddles. For 21.3 and 28.5 MHz this extra inner

Table 7.6 — Examples of half-element taper schedules for three-element Yagis.

14.2 MHz

section	*2a** *(inches)*	x_1 *(inches)*	x_2 *(inches)*	*number of 12 foot lengths*
1	1.125	0	24	1
2	1.000	24	72	3
3	0.875	72	136	3
4	0.750	136	176	2
5	0.625	176	<215	2

21.3 MHz

section	*2a** *(inches)*	x_1 *(inches)*	x_2 *(inches)*	*number of 12 foot lengths*
1	0.875	0	72	3
2	0.750	72	112	2
3	0.625	112	<145	2

28.5 MHz

section	*2a** *(inches)*	x_1 *(inches)*	x_2 *(inches)*	*number of 12 foot lengths*
1	0.875	0	72	3
2	0.750	72	<105	2

Table 7.7 — Radius scaling computations for the three-element Yagi of Table 7.1.

frequency (MHz)	λ (inches)	2a (inches)	K	a (λ)	lengths reflector (λ)	driver (λ)	director (λ)
14.2	831.76	0.875	1901.17	0.00052599	0.49404	0.48572	0.46525
21.3	554.51	0.875	1267.45	0.00078899	0.49366	0.48471	0.46278
28.5	414.42	0.875	947.25	0.00105570	0.49336	0.48389	0.46078

Table 7.8 — Taper calculations for a 14.2 MHz Yagi. Wavelength = 831.76 inches, cylinder radius = 0.00052599 wavelengths, *K* = 1901.17.

reflector

section	$2a^*$ (in.)	x_1 (in.)	x_2 (in.)	K^*	m	θ_2 (degrees)	$f(\theta)$	s_A (in.)
1	1.125	0	24	1478.69	0.95620	10.047	0.97963	22.971
2	1.000	24	72	1663.52	0.97673	30.140	0.74839	47.167
3	0.875	72	136	1901.17	1.00000	56.930	0.04928	64.000
4	0.750	136	176	2218.03	1.02686	73.674	−0.64161	39.334
5	0.625	176	215	2661.64	1.05864	90.000	−0.94675	36.958
								210.430

director

section	$2a^*$ (in.)	x_1 (in.)	x_2 (in.)	K^*	m	θ_2 (degrees)	$f(\theta)$	s_A (in.)
1	1.125	0	24	1478.69	0.95620	10.854	0.97625	22.974
2	1.000	24	72	1663.52	0.97673	32.563	0.70911	47.212
3	0.875	72	136	1901.17	1.00000	61.507	−0.06800	64.000
4	0.750	136	176	2218.03	1.02686	79.598	−0.76544	39.202
5	0.625	176	199	2661.64	1.05864	90.000	−0.97817	21.755
								195.143

section is unnecessary. Table 7.6 shows the specifications for these tapered half-elements where x_1 and x_2 represent the start (inner) and end (outer) positions of each section. Note that the tubing requirements for all three elements are shown in terms of 12 foot lengths.

3) For these three cases it is necessary to scale the original data of Table 7.1 to use the desired average cylinder size. Table 7.7 shows the new scaled cylinder lengths for all three beams using the radius scaling equations (7.2) through (7.6).

4) We are now ready to compute the effect of tapered radii. For the 14.2 MHz elements Table 7.8 shows the flow of calculations; x_1 and x_2 show the start and finish of each section. First a trial guess at the overall reflector half-length is made; I guessed 215 inches in this case. For each section, K^*, m, $f(\theta)$ and s_A are calculated by the technique used to develop Table 7.4. Note

Table 7.9 — Taper calculations for a 21.3 MHz Yagi. Wavelength = 554.51 inches, cylinder radius = 0.00078899 wavelengths, K = 1267.45.

reflector

section	$2a^*$ (in.)	x_1 (in.)	x_2 (in.)	K^*	m	θ_2 (degrees)	$f(\theta)$	s_A (in.)
1	0.875	0	72	1267.45	1.00000	46.286	0.61831	72.000
2	0.750	72	112	1478.69	1.02891	72.000	−0.45812	39.494
3	0.625	112	140	1774.43	1.06309	90.000	−0.93549	26.449
								137.943

director

section	$2a^*$ (in.)	x_1 (in.)	x_2 (in.)	K^*	m	θ_2 (degrees)	$f(\theta)$	s_A (in.)
1	0.875	0	72	1267.45	1.00000	49.846	0.56652	72.000
2	0.750	72	112	1478.69	1.02891	77.538	−0.58380	39.351
3	0.625	112	130	1774.43	1.06309	90.000	−0.96876	16.966
								128.317

Table 7.10 — Taper calculations for a 28.5 MHz Yagi. Wavelength = 414.42 inches, cylinder radius = 0.0010557 wavelengths, K = 947.25.

reflector

section	$2a^*$ (in.)	x_1 (in.)	x_2 (in.)	K^*	m	θ_2 (degrees)	$f(\theta)$	s_A (in.)
1	0.875	0	72	947.25	1.00000	62.308	0.37839	72.000
2	0.750	72	104	1105.13	1.03058	90.000	−0.85138	31.194
								103.194

director

section	$2a^*$ (in.)	x_1 (in.)	x_2 (in.)	K^*	m	θ_2 (degrees)	$f(\theta)$	s_A (in.)
1	0.875	0	72	947.25	1.00000	66.804	0.31051	72.000
2	0.750	72	97	1105.13	1.03058	90.000	−0.89426	24.338
								96.338

that the sum of all cylinder equivalents is 210.43 inches. Next is shown a second trial guess for overall director half-length of 199 inches; the same calculational procedure shows the sum of all cylinder equivalents is 195.143 inches. Using these results and (7.31) and (7.32) we can write:

$$l = 1.04664 l_{eq} - 5.24465 \text{ (14.2 MHz)} \qquad (7.36)$$

which gives the overall tapered half-length, l, for any of the elements in terms of the equivalent cylindrical length l_{eq}, for this particular taper schedule only. Tables 7.9 and 7.10 show exactly the same calculational procedure for the

Table 7.11 — Element half-lengths and positions for the three-element Yagis with taper schedules given in Tables 7.8, 7.9 and 7.10.

		length				position	
freq (MHz)	*element*	*tapered (in.)*	*clamp (in.)*	*boom (in.)*	*total (in.)*	*(λ)*	*(in.)*
14.2	reflector	209.799	0.66	0.09	210.55	0.0	0.0
	driver	206.178	0.66	0.09	206.93	0.15	124.8
	director	197.268	0.66	0.09	198.02	0.30	249.5
21.3	reflector	138.885	0.44	0.06	139.38	0.0	0.0
	driver	136.307	0.44	0.06	136.81	0.15	83.2
	director	129.990	0.44	0.06	130.49	0.30	166.4
28.5	reflector	103.014	0.24	0.06	103.31	0.0	0.0
	driver	101.011	0.24	0.06	101.31	0.15	62.2
	director	96.122	0.24	0.06	96.42	0.30	124.3

21.3 and the 28.5 MHz beam elements: the corresponding equations for tapered half-lengths become:

$$l = 1.03885 l_{eq} - 3.30251 \text{ (21.3 MHz)} \tag{7.37}$$

$$l = 1.02100 l_{eq} - 1.36144 \text{ (28.5 MHz)} \tag{7.38}$$

Application of these equations to the equivalent lengths given in Table 7.7 yields the tapered lengths listed in Table 7.11.

5) We are now ready for the final small boom and boom clamp corrections. For this purpose I assume the elements are U-bolted with saddles to flat plates which in turn are U-bolted with saddles to the boom. Boom diameters are assumed to be three inches for the 14 MHz Yagi and two inches for 21 and 28 MHz. Full plate dimensions are assumed to be six inches wide and eight inches long (14 MHz), five inches wide and six inches long (21 MHz) and four inches wide and four inches long (28 MHz). These plates reduce central pipe inductance and thus cause electrical shortening of the half-element. This shortening amounts to about 0.66 inches (14 MHz), 0.44 inches (21 MHz), and 0.25 inches (28 MHz). Thus, to compensate for the boom clamp each half-element should be lengthened by an equivalent amount; it should be further lengthened by the empirical correction of 1/16 times the boom diameter.

6) With all of these corrections the overall actual physical length of each half-element is shown in Table 7.11. None of these taper schedules is very severe; therefore the actual element lengths are not a great deal longer than the equivalent cylinder lengths; nevertheless the differences are important. Also shown in Table 7.11 is the position along the boom for each element, expressed both in wavelengths and in inches. Although I have not experimentally tested any of these particular three-element beams I am

confident that their performance will be excellent, and moreover they all should be easy to construct.

Summary

A practical Yagi antenna design requires the selection of a suitable cylindrical element design, such as one of those developed in Chapters 2 and 3, followed by corrections to element lengths to accommodate practical element construction and mounting techniques. First one must scale the cylindrical element to an equivalent cylinder of the desired radius. Next, the overall element length must be altered to take into account the desired element taper schedule. Finally, some small length corrections are needed to compensate for boom mounting plates and other mechanical hardware.

NOTES:

[1]Viezbicke, P. "Yagi Antenna Design," *NBS Technical Note 688,* U.S. Department of Commerce, Washington, D.C., December, 1976.

[2]Uda, S., and Mushiake, Y., *Yagi-Uda Antenna*, Research Institute of Electrical Communication, Tohoku University, Sendai, Japan: Sasaki Ltd., 1954.

CHAPTER 8

PRACTICAL AMATEUR YAGI ANTENNAS

In this chapter I will provide examples of practical Yagi designs that should perform quite well in each of the Amateur bands, 7 MHz, 14 MHz, 21 MHz and 28 MHz. At the lower frequencies mechanical considerations dominate the design and only rather simple Yagi electrical designs are worth considering; whereas, at the higher frequencies the mechanical obstacles are far easier to overcome and permit more sophisticated electrical design. In previous chapters it has been shown that a Yagi antenna is somewhat dependent on its immediate environment; this is unfortunate since it means that the "best" design is not best for all applications. Nevertheless a good design for free space is actually quite good for most applications—particularly where the antenna is not too close to ground or other similar antennas. Therefore I shall show only free space performance and will make no attempt to adjust for nearby structures nor will I attempt any element positioning adjustments to optimize the F/B ratio as described in Chapter 3.

For each Yagi design there are a number of sequential steps required to compute and present useful information. The first step is to choose a boom length; this choice is based on three considerations. First, there are preferred lengths; recall that for a two-element beam the boom should be about 0.15 wavelengths long and for all others (for best *F/B*) it should be near an odd multiple of one-quarter wavelength. Second, it is nice to construct the boom from a "standard" length or lengths of strong aluminum tubing, leaving a little room at the ends for element mounting hardware. Third, there is some gain advantage to longer rather than shorter booms. These three considerations permit a rational selection of boom length.

The next step is to choose the element radius, that is, the average radius at the center of tapered half-element; the lower limit of radius is determined by mechanical strength requirements. Larger values of radius will improve the bandwidth of the system but will certainly make the antenna heavy and more difficult to support and turn in the wind.

Next is a choice of number of elements and boom placement. It has been shown that evenly spaced elements on the boom give performance basically as good as that with any other arrangement and I would generally favor such a placement. I also generally favor a choice of an even number of elements just because at the boom center there is no element to interfere with mounting the antenna to the mast or rotor system. However a three-element beam is a special case; its performance for boom lengths approximating one-quarter wavelength is as good as four elements on the same boom and has the advantages of excellent accessibility to the driver matching system (from the fixed mast) and smaller wind load than the equivalent four-element beam. For this special case the disadvantage of driver element interference with the mounting system can be overcome by moving the driver along the boom a short distance; ultimate beam performance is essentially unaffected.

Having chosen a boom length and number of elements I start with the closest preferred Yagi design, such as one of those shown in Table 7.1; this design is then scaled to the new selected radius, to obtain preliminary element lengths. Using the chosen boom positions and the chosen radius and the preliminary values of length, the performance is computed. It will be found that the frequency of maximum F/B will generally not be at the design frequency ($F = 1$), but at F_F. Now readjust all elements by a factor F_F and recompute; it will be found that the frequency of maximum F/B will now be close to the design frequency. Perhaps a second iteration will be needed to make the agreement precise.

Now it will be found that the frequency of zero driving-point reactance is generally not $F = 1$ but at F_X; readjust the driver element length only by a factor F_X and recompute. Again a second iteration may be needed to make the driving-point reactance zero at $F = 1$. This is the final design and computed performance can now be recorded and displayed in several ways. I shall, for each case, show a table listing, for a series of frequencies, the free space gain, F/B ratio, and driving-point resistance and reactance. The gain and F/B ratio will also be shown on a plot for easy visualization. A second diagram will show both E- and H-plane radiation patterns.

For each case it is necessary to design the actual tapered element. Choose convenient lengths of telescoping tubing and devise a taper schedule which is mechanically adequate and where the remaining variable is only the length of the outermost section. Of course length corrections should be made for both mounting hardware and boom size. Perhaps the best way to make these corrections is to compute the effective diameter of the mounting plate and boom combination and use this as the initial short element section of the taper schedule for a half-element. This initial short section has an effective length equal to the physical half-length of the mounting plate minus the small boom correction (6 percent of the boom radius). Thus by the same technique as that shown in Chapter 7, the entire (corrected) element taper schedule can produce an equation relating the physical length of the fully corrected tapered half-element, l, to the equivalent cylindrical half-length, l_{eq}.

The design and performance tables, performance plots and radiation patterns for each of the Yagis described here are grouped together at the end of this chapter. For convenience, the tables and figures for the various antennas designs are summarized here:

		Tables			*Figures*	
antenna	*page no.*	*taper table*	*design table*	*performance table*	*performance plot*	*radiation patterns*
7 MHz	8-10	8.1				
2-el	8-12		8.2	8.3	8.1	8.2
3-el			8.4	8.5	8.3	8.4
14 MHz		8.6				
3-el	8-14		8.7	8.8	8.5	8.6
4-el	8-16		8.9	8.10	8.7	8.8
4-el	8-18		8.11	8.12	8.9	8.10
5-el	8-20		8.13	8.14	8.11	8.12
6-el	8-22		8.15	8.16	8.13	8.14
21 MHz		8.17				
3-el	8-24		8.18	8.19	8.15	8.16
6-el	8-26		8.20	8.21	8.17	8.18
28 MHz		8.22				
3-el	8-28		8.23	8.24	8.19	8.20
6-el	8-30		8.25	8.26	8.21	8.22
8-el	8-32		8.27	8.28	8.23	8.24

Designs For 7.15 MHz

I choose a taper schedule starting each half-element with a two inch diameter tube 20 feet long. At its outer end I use a one inch diameter tube 12 feet long which is supported from a pair of machined reducers in the two inch tube. The overlap region is made one foot long to give adequate strength. The outer end of the one inch diameter tube supports an inserted 7/8 inch diameter telescoping tube long enough to complete the half-element.

I choose a boom diameter of three inches which results in a virtually negligible shortening of the initial (mounting plate) section of 0.09 inches. I next choose a central element mounting plate 6 by 24 by 1/2 inches; such a plate allows room for two U-bolts and saddles on each half-element and two larger U-bolts and saddles to clamp to the boom. The (half) mounting plate and two inch clamped element becomes the first section in the element taper schedule. The effective radius is calculated as shown in Chapter 7, and the result is 1.567 inches. The complete taper schedule is shown in Table 8.1, and by the methods shown in Chapter 7 produces:

$$l = 1.06793 l_{eq} - 11.68745 \tag{8.1}$$

Two-element Beam

For a two-element beam a convenient choice for a boom is a three inch diameter tube 20 feet long; it is close to the desired 0.15 wavelength. Some room is needed at each end for the mounting plates of the end element; I shall assume four inches. I choose the element diameter as two inches diameter which experience shows is mechanically strong enough and does not sag excessively. Using these chosen values and adjusting element lengths as described in Chapter 7 I arrive at the final design shown in Table 8.2. Performance over the 7 MHz band is shown in Table 8.3 and plots of gain and *F/B* ratio are shown in Fig. 8.1. Fig. 8.2 shows the angular performance of the beam in free space. Note that the 3 dB total beamwidth is 119 degrees for the *H*-plane and 74 degrees for the *E*-plane.

By using (8.1) and the design data in Table 8.2 we can now specify the real two-element beam in terms of boom positions and the overall tapered element length; the result is included in Table 8.2. Note that the actual lengths of the tapered elements are considerably longer than the equivalent cylindrical lengths.

Three-element Beam

For this case I choose a boom three inches in diameter but 40 feet long, again leaving four inches at each end for element mounting plates. I choose the same element specifications as above, including the taper schedule. However I shall move the driven element 24 inches off the boom center to clear interference with mounting to rotor or mast. Using these values the final adjusted (cylindrical elements) design is shown in Table 8.4 and performance over the 7 MHz band is shown in Table 8.5. Plots of gain and *F/B* ratio are shown in Fig. 8.3 and the angular performance of the beam is shown in Fig. 8.4. The 3 dB total beamwidth is 100 degrees for the *H*-plane and 68 degrees for the *E*-plane.

I choose the same element taper schedule and mounting plate information as selected for the two-element design; therefore (8.1) is valid and leads directly to the final beam specification shown in Table 8.4. Note that the element lengths are somewhat different than those for the two-element beam.

Designs For 14.2 MHz

I choose a basic cylindrical element 7/8 inches in diameter which experience shows is large enough for practical 14 MHz elements. Table 8.6 shows two different taper schedules which can be used; the first is a heavily tapered half-element in common use, and the second is a lightly tapered half-element that can be easily made from standard telescoping 6061-T6 aluminim seamless tubing with 0.058 inch wall thickness. In both cases the taper schedules include an initial short section consisting of the mounting plate clamped by U-bolts and saddles to the initial element tubing and whose equivalent radius is computed as shown in Chapter 7. For both schedules I have assumed an aluminum mounting plate whose full dimensions are eight inches long, six inches wide, and 3/8 inches thick. The effective length of this initial section for the half-element is therefore four inches minus the small correction for boom diameter; in Table 8.6 I have assumed a boom which is three inches in diameter. The boom correction is so small that these half-element schedules should apply quite well to any boom size from two inches to four inches.

These two schedules through the methods explained in Chapter 7 produce useful equations relating the cylindrical and tapered half-lengths. For the heavy taper:

$$l = 1.06938 l_{eq} - 5.09785 \tag{8.2}$$

and for the light taper:

$$l = 1.04664 l_{eq} - 4.57166 \tag{8.3}$$

Three-element Beam

A good choice for a boom is a tube 20 feet long with about four inches at each end used for element mounting hardware. To help clear mechanical interference between the driven element and the rotor or mast mounting bracket, I will move the driver 12 inches off center. With these boom positions and the initial element lengths shown in Table 7.1, the procedure explained in Chapter 7 produces the final design shown in Table 8.7. Performance over the 14 MHz band is shown in Table 8.8 and plots of gain and *F/B* ratio are shown in Fig. 8.5. Fig. 8.6 shows the angular performance of the beam in free space. The 3 dB total beamwidth is 100 degrees for the *H*-plane and 68 degrees for the *E*-plane.

Using the two taper schedules of Table 8.6 and the resulting equations (8.2) and (8.3) with the basic beam data of Table 8.7 we can now give complete specifications for the three-element beam; these are included in Table 8.7. Note that the tapered element lengths are quite different for the two different taper schedules.

Four-element Short-boom Beam

Exactly the same procedure can be used to design a four-element beam on the same 20 foot boom. Since there is no element mechanical interference at the center of the boom I shall space elements equally. The final design for such a beam is shown in Table 8.9 and performance over the 14 MHz band is shown in Table 8.10. Plots of gain and *F/B* ratio are shown in Fig. 8.7 and angular performance of the beam is shown in Fig. 8.8. The 3 dB total beamwidth is 88 degrees for the *H*-plane and 64 degrees for the *E*-plane. Note from Table 8.10 that this four-element beam does not show a very high *F/B* ratio; the reason is demonstrated in Fig. 8.8 where it can be seen that there is a small back lobe between minima at angles of ± 150 degrees. This back lobe arises because the boom is a bit too long; it can be reduced by shortening the boom but then the peak gain would be reduced as well. The beam is actually a pretty good beam generating rather low back radiation over a substantial range in angle.

Four-element Long Beam

One could use four elements on a somewhat longer boom which results in larger gain, but at a very significant loss in *F/B* ratio. However, we know that if the boom is lengthened to about 3/4 wavelengths the *F/B* ratio improves again. It is interesting to see if a beam with such a long boom with only four elements is a good performer.

I shall choose a boom of a convenient total length, 54 feet, and diameter three inches, and again allow four inches at each end for element mounting hardware. With elements equally spaced along the boom the final design for this long four-element beam is shown in Table 8.11. The beam performance over the 14 MHz band is shown in Table 8.12 and plots of gain and *F/B* ratio are shown in Fig. 8.9. Fig. 8.10 shows the angular performance of the beam in free space. The 3 dB total beamwidth is 63 degrees for the *H*-plane and 53 degrees for the *E*-plane. Observe that the performance is indeed excellent, and as we shall shortly see, is essentially the same as a five or a six-element implementation on the same boom.

Five-element Beam

For this case I choose the same boom as for the long-boom four-element design, but will equally space five elements along the boom; the design is shown in Table 8.13. The beam performance over the 14 MHz band is shown in Table 8.14 and plots of gain and *F/B* ratio are shown in Fig. 8.11. Figure 8.12 shows the free space angular performance; the 3 dB total beamwidth is 66 degrees for the *H*-plane and 54 degrees for the *E*-plane.

Six-element Beam

Using exactly the same technique as for the five-element beam the complete design for an equally-spaced six-element beam on the same 54 foot boom is shown in Table 8.15, and the performance of the beam over the 14 MHz band is shown in Table 8.16. Plots of gain and *F/B* performance are shown in Fig. 8.13 and of angular performance are shown in Fig. 8.15. Full 3 dB beamwidths are 63 degrees (*H*-plane) and 53 degrees (*E*-plane).

Note that there is not a great deal of difference between the performances of the four, five and six-element beams when constructed on the same 54-foot boom. Variations in peak *F/B* probably are not very significant; one can in principle improve the already excellent figures by optimizing the exact placement of elements along the boom. The gain from the four-element beam, compared to the others, is a bit more sensitive to frequency variations, and—not indicated on these figures — is also more likely to be affected adversely by nearby scatterers. Both the five and six-element versions are fine performers either alone or in stacked systems; my own preference is the six-element version since it can be mounted at its center with no mechanical interference from any element.

Designs For 21.3 MHz

I choose a basic cylindrical element 7/8 inches in diameter with the taper schedule shown in Table 8.17. Two initial (half-length) sections are shown; the first caused by the central mounting plate which I have chosen to be six inches long, five inches wide, and 3/8 inches thick. There is a small boom correction; for a three inch boom it is 0.09 inches. The second initial section is a short piece of one inch diameter telescoping tubing for reinforcing purposes only. This taper schedule results in the equation:

$$l = 1.03875 l_{eq} - 2.59375 \qquad (8.4)$$

Three-element Beam

In exactly the same way as the 14 MHz beams were specified, a three-element 21 MHz beam can be derived. The final design is shown in Table 8.18 and its performance over the 21 MHz band shown in Table 8.19. Fig. 8.15 shows plots of the gain and *F/B* performance over the band and Fig. 8.16 shows the free space angular performance. Full 3 dB beamwidths are 96 degrees (*H*-plane) and 67 degrees (*E*-plane).

Six-element Beam

Following the same procedure a six-element 21 MHz beam can be specified. Table 8.20 shows the design and Table 8.21 the performance over the 21 MHz band. Fig. 8.17 shows plots of gain and *F/B* over the band and Fig. 8.18 shows free space angular performance. Full 3 dB beamwidths are 63 degrees (*H*-plane) and 53 degrees (*E*-plane).

Designs For 28.5 MHz

I again choose a basic cylindrical element 7/8 inches in diameter with a taper schedule shown in Table 8.22. Two initial (half-length) sections are shown; the first caused by the central mounting plate, which I have chosen to be four inches square, and 1/4 inches thick. I choose a boom two inches diameter resulting in a small correction of 0.06 inches. A second initial section is a short piece of reinforcing one inch diameter tubing. This taper schedule results in the taper equation:

$$l = 1.02100 l_{eq} - 0.96835 \qquad (8.5)$$

Three-element Beam

A three-element 28 MHz beam can be derived by the same technique used above. The result is shown in Table 8.23 and its performance over the 28 MHz band shown in Table 8.24. Fig. 8.19 shows plots of gain and *F/B* performance over the band and Fig. 8.20 shows the free space angular performance. Full 3 dB beamwidths are 98 degrees (*H*-plane) and 68 degrees (*E*-plane).

Six-element Beam

A six-element 28 MHz beam can be similarly specified. Table 8.25 shows the final design and Table 8.26 the performance over the 28 MHz band. Fig. 8.21 shows plots of gain and *F/B* over the 28 MHz band and Fig. 8.22 shows free space angular performance. Full 3 dB beamwidths are 63 degrees (*H*-plane) and 53 degrees (*E*-plane).

Eight-element Beam

In the same way a long eight-element 28 MHz beam can be specified. Table 8.27 shows this design, and Table 8.28 shows the performance over the 28 MHz band. Figure 8.23 shows plots of gain and *F/B* over the 28 MHz band and Fig. 8.24 shows free space angular performance. Full 3 dB beamwidths are 53 degrees (*H*-plane) and 47 degrees (*E*-plane).

Summary

A number of practical Yagi designs have been developed for the Amateur Radio bands: 7, 14, 21 and 28 MHz. Theoretical performance in free space is shown for each design. A substantial number of tables and figures have been prepared.

Although these designs are all theoretical, I believe they should give excellent performance in practice. Minor modifications might be needed to compensate for electrical interaction with ground or with other adjacent antennas; such modifications will probably be inconsequential if the Yagi is sufficiently high about ground (see Chapter 5) and if it is sufficiently separated from adjacent antennas (Chapter 6).

Design and Performance Information for a Two-Element 7 MHz Yagi

Table 8.1 — Taper schedule for 7 MHz elements.

section	length (inches)	diameter (inches)	overlap (inches)
0	11.91	3.134	
1	228.0	2.000	12
2	132.0	1.000	12
3	variable	0.875	>8

Table 8.2 — Element positions and lengths for a two-element 7 MHz Yagi. Element diameters are two inches (radius = 0.00060537 wavelengths).

	position		l_{eq}		l
element	(inches)	(λ)	(inches)	(λ)	(inches)
reflector	0	0.0	815.24	0.49352	847.24
driver	232	0.1405	776.44	0.47003	805.81

Table 8.3 — Performance of a two-element 7 MHz Yagi.

freq. (MHz)	gain (dBi)	F/B (dB)	resistance (ohms)	reactance (ohms)
6.80	1.35	−5.6	23.6	−89.4
6.85	3.18	−3.7	21.2	−76.6
6.90	4.73	−1.7	19.9	−63.3
6.95	5.88	0.3	19.7	−49.8
7.00	6.59	2.3	20.8	−36.4
7.05	6.94	4.2	22.9	−23.5
7.10	7.02	6.1	26.0	−11.4
7.15	6.93	7.8	29.8	−0.3
7.20	6.76	9.2	33.8	9.9
7.25	6.56	10.3	37.4	19.2
7.30	6.34	10.8	41.9	27.6
7.35	6.12	10.9	45.6	35.4
7.40	5.92	10.6	49.0	42.6
7.45	5.73	10.1	52.1	49.4
7.50	5.55	9.6	54.8	55.9

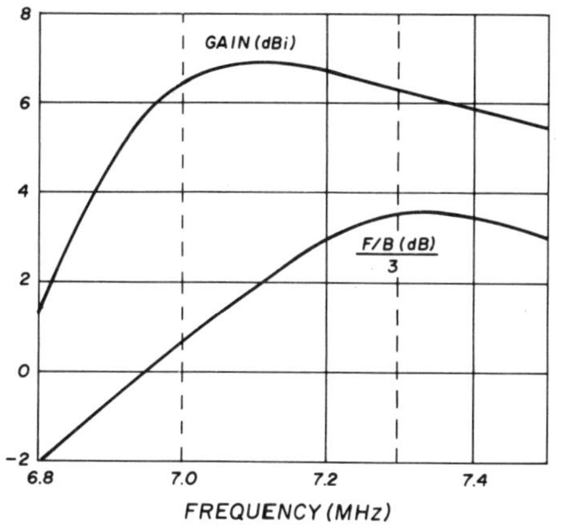

Fig. 8.1 — Free space performance of a two-element 7 MHz beam.

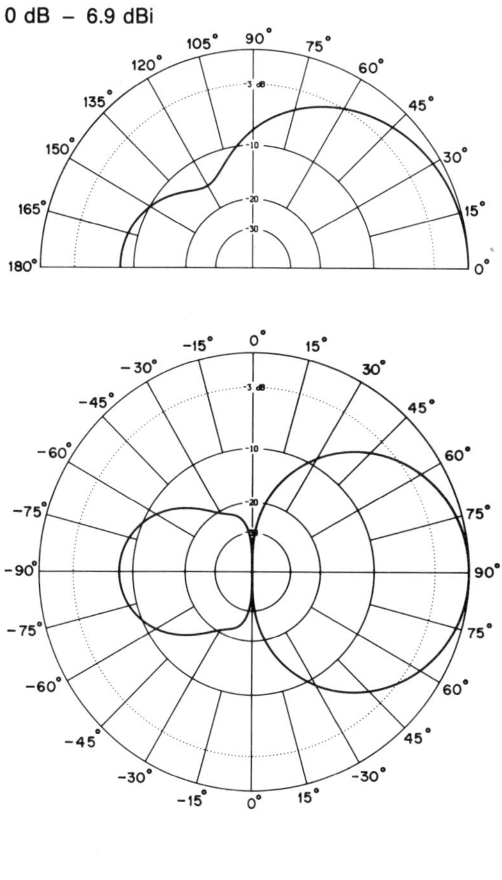

Fig. 8.2 — Two-element beam *H*-plane and *E*-plane patterns at 7.15 MHz.

Design and Performance Information for a Three-Element 7 MHz Yagi

Table 8.4 — Element positions and lengths for a three-element 7 MHz Yagi. Element diameters are two inches (radius = 0.00060537 wavelengths).

	position		l_{eq}		l
element	*(inches)*	*(λ)*	*(inches)*	*(λ)*	*(inches)*
reflector	0	0.0	819.20	0.49589	851.47
driver	260	0.15740	807.01	0.48854	838.45
director	472	0.28573	770.26	0.46629	799.21

Table 8.5 — Performance of a three-element 7 MHz Yagi.

freq. (MHz)	*gain (dBi)*	*F/B (dB)*	*resistance (ohms)*	*reactance (ohms)*
6.80	5.64	2.7	30.2	−55.2
6.85	6.39	4.8	30.6	−46.4
6.90	6.91	7.1	30.7	−38.6
6.95	7.25	9.6	30.0	−31.5
7.00	7.47	12.5	28.3	−24.5
7.05	7.64	16.2	25.6	−17.1
7.10	7.80	21.8	22.3	−9.0
7.15	7.99	29.9	18.7	0.0
7.20	8.21	21.4	15.2	9.9
7.25	8.46	15.4	12.0	20.7
7.30	8.69	11.4	9.3	32.2
7.35	8.79	8.1	7.3	44.2
7.40	8.56	5.3	6.0	56.7
7.45	7.76	2.8	5.7	69.3
7.50	6.36	0.6	6.3	81.8

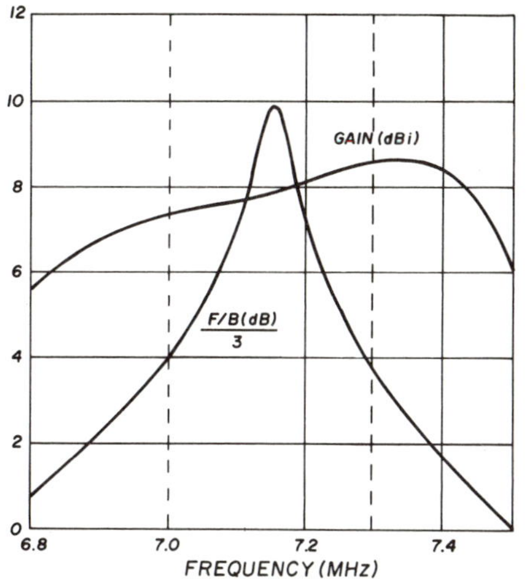

Fig. 8.3 — Free space performance of a three-element 7 MHz beam.

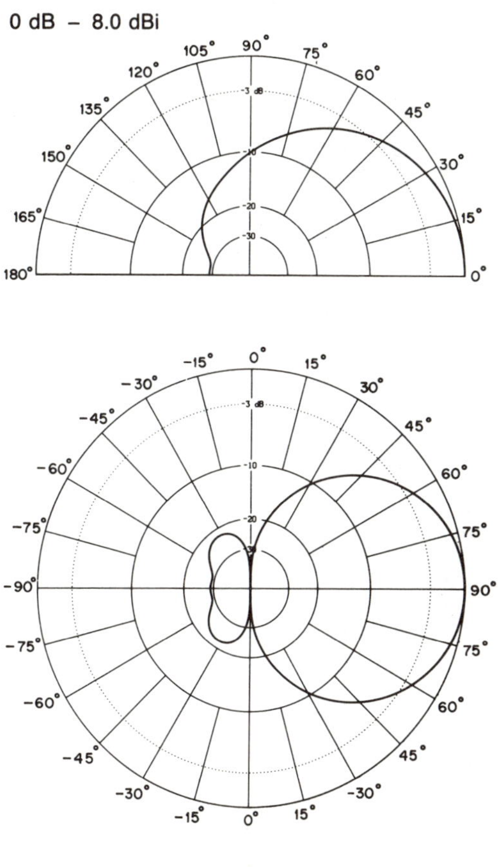

Fig. 8.4 — Three-element beam *H*-plane and *E*-plane patterns at 7.15 MHz.

Design and Performance Information for a Three-Element 14 MHz Yagi

Table 8.6 — Taper schedule for 14 MHz elements.

	heavy taper		*light taper*	
section	*length (inches)*	*diameter (inches)*	*length (inches)*	*diameter (inches)*
0	3.91	2.577	3.91	2.548
1	32.0	1.250	20.0	1.125
2	50.0	1.125	48.0	1.000
3	44.0	0.875	64.0	0.875
4	32.0	0.625	40.0	0.750
5	variable	0.500	variable	0.625

Table 8.7 — Element positions and lengths for a three-element 14 MHz Yagi. Element diameters are 7/8 inches (radius = 0.00052599 wavelengths).

	position		l_{eq}		l_H	l_L
element	*(inches)*	*(λ)*	*(inches)*	*(λ)*	*(inches)*	*(inches)*
reflector	0	0.0	413.02	0.49656	431.48	423.14
driver	128	0.15389	407.01	0.48933	425.05	416.85
director	232	0.27893	388.96	0.46763	405.75	397.96

Table 8.8 — Performance of a three-element 14 MHz Yagi.

freq. (MHz)	*gain (dBi)*	*F/B (dB)*	*resistance (ohms)*	*reactance (ohms)*
13.80	7.21	9.2	30.5	−31.7
13.85	7.33	10.6	29.8	−28.9
13.90	7.43	12.2	28.8	−24.8
13.95	7.51	13.9	27.5	−21.8
14.00	7.59	15.8	25.9	−17.5
14.05	7.67	18.1	24.2	−14.1
14.10	7.75	21.1	22.3	−9.3
14.15	7.84	24.8	20.4	−5.4
14.20	7.94	27.3	18.5	−0.0
14.25	8.04	24.5	16.6	5.5
14.30	8.16	20.6	14.8	10.2
14.35	8.29	17.5	13.0	16.2
14.40	8.42	14.9	11.4	21.4
14.45	8.54	12.7	9.9	27.8
14.50	8.65	10.8	8.6	33.3
14.55	8.72	9.1	7.5	40.0
14.60	8.72	7.5	6.7	45.7

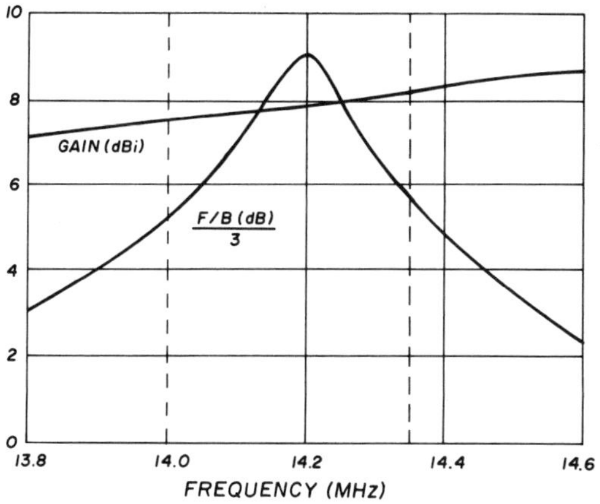

Fig. 8.5 — Free space performance of a three-element 14 MHz beam.

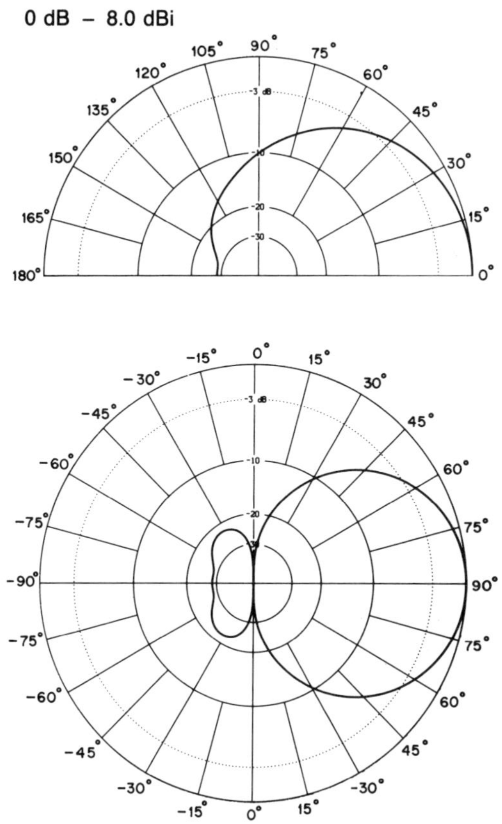

Fig. 8.6 — Three-element beam *H*-plane and *E*-plane patterns at 14.2 MHz.

Design and Performance Information for a Short-Boom Four-Element 14 MHz Yagi

Table 8.9 — Element positions and lengths for a short-boom four-element 14 MHz Yagi. Element diameters are 7/8 inches (radius = 0.00052599 wavelengths).

	position		l_{eq}		l_H	l_L
element	*(inches)*	*(λ)*	*(inches)*	*(λ)*	*(inches)*	*(inches)*
reflector	0.0	0.0	411.66	0.49493	430.03	421.72
driver	77.33	0.09297	404.73	0.48659	422.61	414.46
D1	154.67	0.18595	387.68	0.46609	404.38	396.62
D2	232.00	0.27893	387.68	0.46609	404.38	396.62

Table 8.10 — Performance of a short-boom four-element 14 MHz Yagi.

freq. (MHz)	*gain (dBi)*	*F/B (dB)*	*resistance (ohms)*	*reactance (ohms)*
13.80	2.79	−0.7	9.3	−46.4
13.85	4.05	1.5	8.5	−41.0
13.90	5.10	3.7	8.0	−34.5
13.95	5.95	6.0	7.5	−29.2
14.00	6.64	8.4	7.2	−22.9
14.05	7.19	11.1	6.9	−17.7
14.10	7.64	14.0	6.7	−11.4
14.15	7.99	17.0	6.6	−6.2
14.20	8.27	18.5	6.5	0.1
14.25	8.49	17.1	6.5	6.4
14.30	8.64	14.8	6.6	11.7
14.35	8.74	12.7	6.8	18.2
14.40	8.78	11.0	7.1	23.7
14.45	8.78	9.5	7.7	30.6
14.50	8.74	8.4	8.5	36.7
14.55	8.66	7.4	9.9	44.3
14.60	8.55	6.7	12.2	51.7

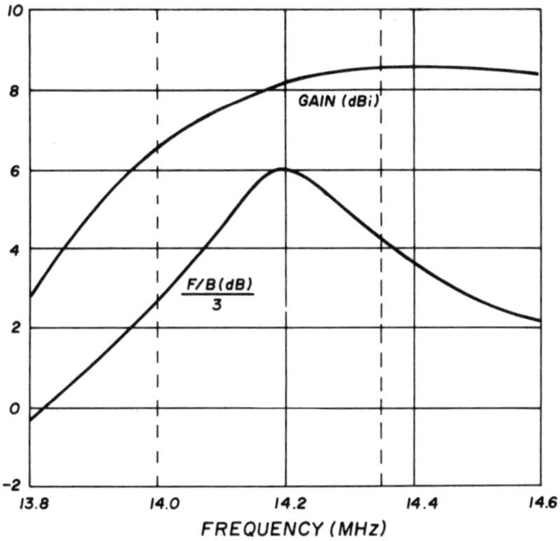

Fig. 8.7 — Free space performance of a short-boom four-element 14 MHz beam.

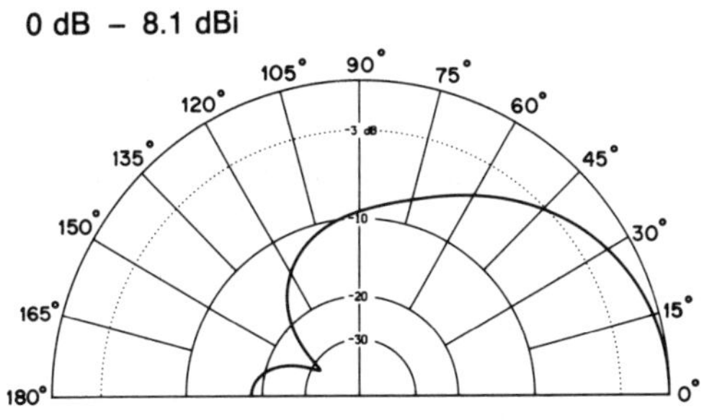

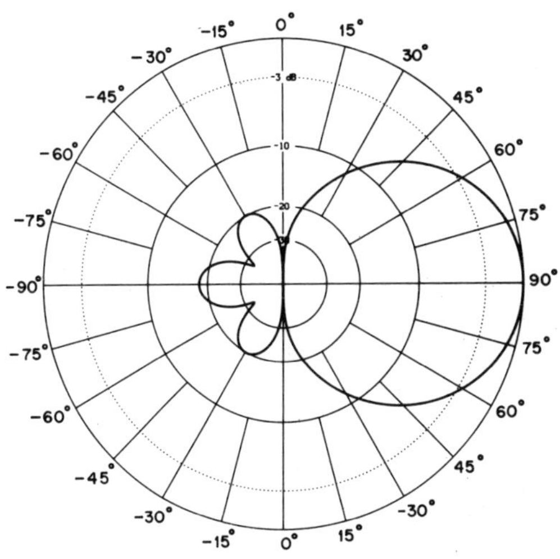

Fig. 8.8 — Short-boom four-element beam *H*-plane and *E*-plane patterns at 14.2 MHz.

Design and Performance Information for a Long-Boom Four-Element 14 MHz Yagi

Table 8.11 — Element positions and lengths for a long-boom four-element 14 MHz Yagi. Element diameters are 7/8 inches (radius = 0.00052599 wavelengths).

	position		*l_{eq}*		*l_H*	*l_L*
element	*(inches)*	*(λ)*	*(inches)*	*(λ)*	*(inches)*	*(inches)*
reflector	0.0	0.0	407.26	0.48964	425.32	417.11
driver	213.33	0.25648	397.54	0.47795	414.93	406.94
D1	426.67	0.51297	383.54	0.46112	399.85	392.28
D2	640.00	0.76945	383.54	0.46112	399.95	392.28

Table 8.12 — Performance of a long-boom four-element 14 MHz Yagi.

freq. (MHz)	*gain* (dBi)	*F/B* (dB)	*resistance* (ohms)	*reactance* (ohms)
13.80	9.37	8.4	50.4	−27.9
13.85	9.51	9.4	50.0	−25.8
13.90	9.64	10.5	49.1	−22.6
13.95	9.77	11.9	47.6	−20.6
14.00	9.91	13.6	45.4	−17.3
14.05	10.06	15.8	42.6	−14.6
14.10	10.23	18.9	39.2	−10.3
14.15	10.43	24.2	35.3	−6.3
14.20	10.64	34.2	31.2	−0.2
14.25	10.86	23.5	27.1	7.0
14.30	11.07	17.3	23.2	14.2
14.35	11.21	13.3	20.0	23.7
14.40	11.18	10.2	17.5	33.0
14.45	10.87	7.7	16.3	44.2
14.50	10.16	5.5	16.4	54.7
14.55	9.06	3.6	18.1	66.6
14.60	7.68	1.9	21.2	77.0

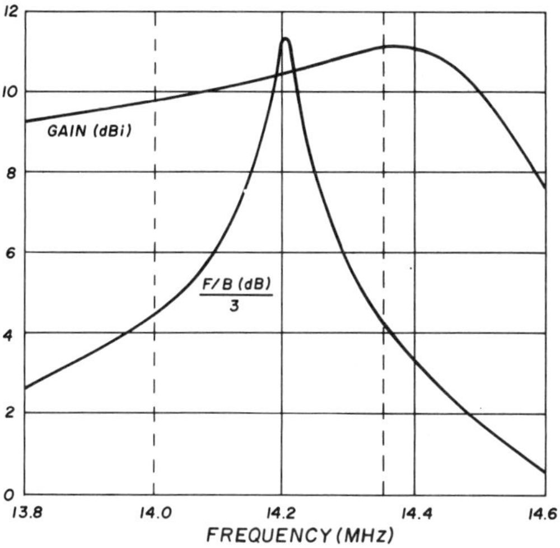

Fig. 8.9 — Free space performance of a long-boom four-element 14 MHz beam.

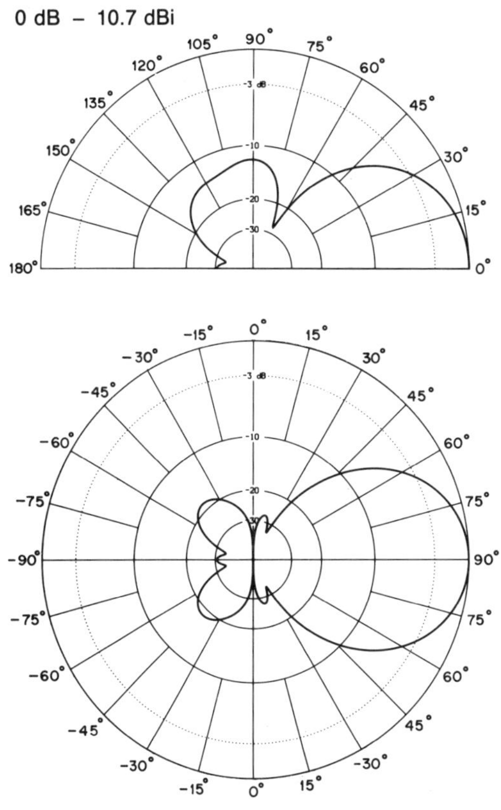

Fig. 8.10 — Long-boom four-element beam *H*-plane and *E*-plane patterns at 14.2 MHz.

Design and Performance Information for a Five-Element 14 MHz Yagi

Table 8.13 — Element positions and lengths for a five-element 14 MHz Yagi. Element diameters are 7/8 inches (radius = 0.00052599 wavelengths).

	position		l_{eq}		l_H	l_L
element	*(inches)*	*(λ)*	*(inches)*	*(λ)*	*(inches)*	*(inches)*
reflector	0	0.0	414.00	0.49774	432.53	424.17
driver	160	0.19236	398.66	0.47930	416.12	408.11
D1	320	0.38473	374.57	0.45033	390.36	382.90
D2	480	0.57709	374.57	0.45033	390.36	382.90
D3	640	0.76945	374.57	0.45033	390.36	382.90

Table 8.14 — Performance of a five-element 14 MHz Yagi.

freq. (MHz)	*gain (dBi)*	*F/B (dB)*	*resistance (ohms)*	*reactance (ohms)*
13.80	9.79	10.6	38.5	−31.2
13.85	9.90	11.9	38.5	−28.2
13.90	9.99	13.3	38.3	−24.1
13.95	10.08	14.9	37.8	−21.1
14.00	10.17	16.9	37.0	−16.8
14.05	10.25	19.4	35.9	−13.5
14.10	10.34	22.9	34.7	−8.9
14.15	10.43	28.7	33.3	−5.2
14.20	10.52	46.5	31.9	−0.0
14.25	10.61	28.4	30.4	5.5
14.30	10.69	22.4	29.0	10.2
14.35	10.77	18.7	27.7	16.4
14.40	10.83	16.1	26.5	21.8
14.45	10.86	14.0	25.6	28.6
14.50	10.85	12.3	25.0	34.7
14.55	10.79	10.8	24.9	42.3
14.60	10.68	9.5	25.3	49.0

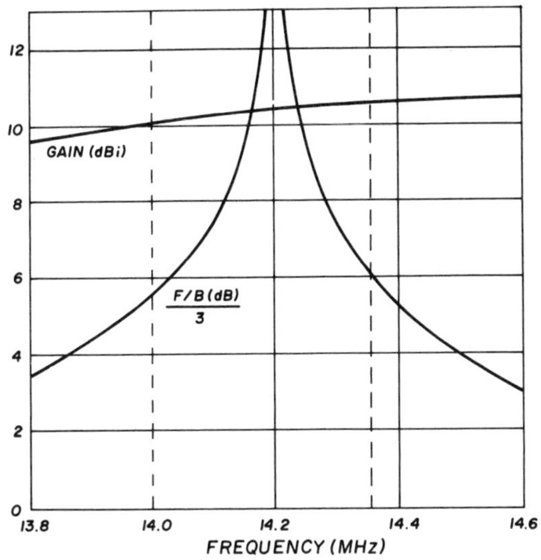

Fig. 8.11 — Free space performance of a five-element 14 MHz beam.

0 dB - 10.5 dBi

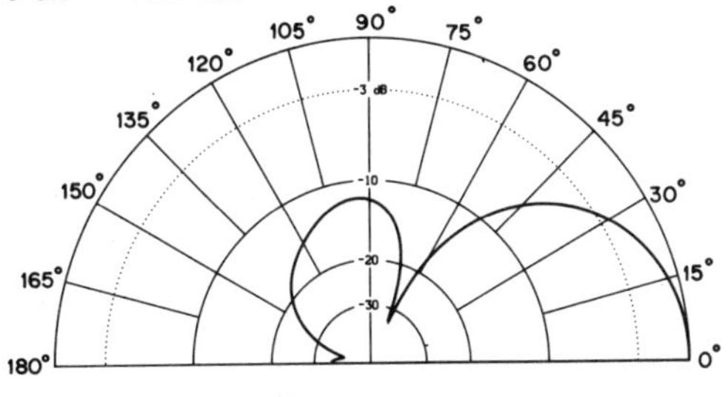

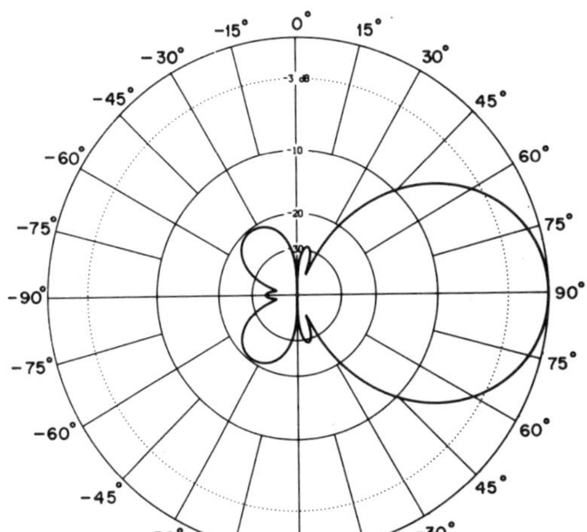

Fig. 8.12 — Five-element beam *H*-plane and *E*-plane patterns at 14.2 MHz.

Design and Performance Information for a Six-Element 14 MHz Yagi

Table 8.15 — Element positions and lengths for a six-element 14 MHz Yagi. Element diameters are 7/8 inches (radius = 0.00052599 wavelengths).

element	position (inches)	position (λ)	l_{eq} (inches)	l_{eq} (λ)	l_H (inches)	l_L (inches)
reflector	0	0.0	410.47	0.49349	428.75	420.47
driver	128	0.15389	398.96	0.47966	416.44	408.42
D1	256	0.30778	371.37	0.44649	386.94	379.55
D2	384	0.46167	371.37	0.44649	386.94	379.55
D3	512	0.61556	371.37	0.44649	386.94	379.55
D4	640	0.76945	371.37	0.44649	386.94	379.55

Table 8.16 — Performance of a six-element 14 MHz Yagi.

freq. (MHz)	gain (dBi)	F/B (dB)	resistance (ohms)	reactance (ohms)
13.80	9.29	8.3	24.0	−41.4
13.85	9.58	9.7	23.7	−36.9
13.90	9.83	11.3	23.5	−31.3
13.95	10.05	13.0	23.2	−26.8
14.00	10.23	15.1	22.9	−21.2
14.05	10.39	17.6	22.5	−16.7
14.10	10.54	20.9	22.1	−11.0
14.15	10.66	25.5	21.7	−6.2
14.20	10.77	29.6	21.3	−0.2
14.25	10.87	25.6	21.0	5.9
14.30	10.94	21.2	20.9	11.1
14.35	10.99	18.1	20.8	17.6
14.40	11.01	15.8	21.0	23.2
14.45	11.01	13.9	21.5	30.0
14.50	10.97	12.5	22.4	36.0
14.55	10.89	11.2	23.7	43.2
14.60	10.78	10.1	25.6	49.6

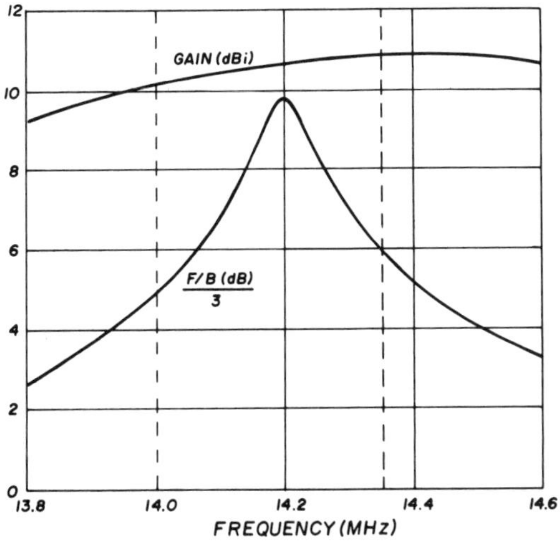

Fig. 8.13 — Free space performance of a six-element 14 MHz beam.

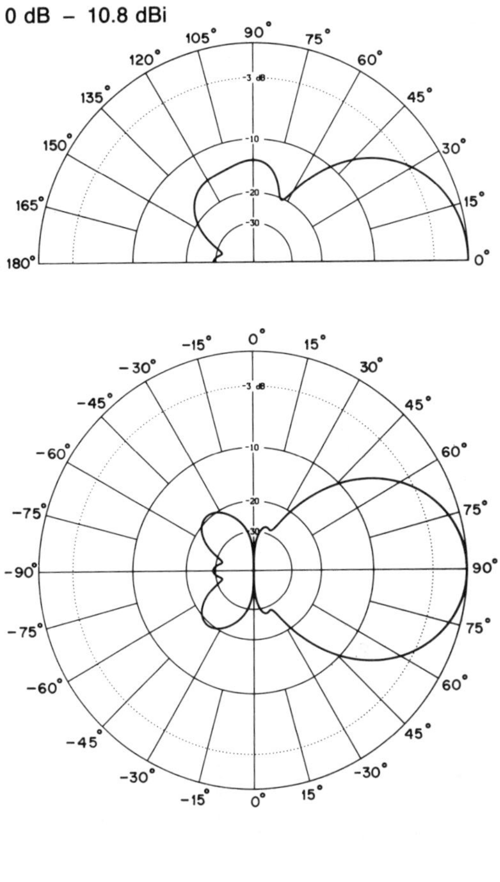

Fig. 8.14 — Six-element beam *H*-plane and *E*-plane patterns at 14.2 MHz.

Design and Performance Information for a Three-Element 21 MHz Yagi

Table 8.17 — Taper schedule for 21 MHz elements.

section	*length (inches)*	*diameter (inches)*
00	2.91	2.206
0	3.0	1.000
1	66.0	0.875
2	40.0	0.750
3	variable	0.625

Table 8.18 — Element positions and lengths for a three-element 21 MHz Yagi. Element diameters are 7/8 inches (radius = 0.00078899 wavelengths).

	position		l_{eq}		l
element	*(inches)*	*(λ)*	*(inches)*	*(λ)*	*(inches)*
reflector	0	0.0	274.08	0.49427	279.51
driver	81	0.14608	269.31	0.48567	274.56
director	162	0.29215	256.93	0.46335	261.70

Table 8.19 — Performance of a three-element 21 MHz Yagi.

freq. (MHz)	*gain (dBi)*	*F/B (dB)*	*resistance (ohms)*	*reactance (ohms)*
20.80	7.38	10.5	22.7	−26.6
20.85	7.48	11.6	22.3	−24.3
20.90	7.57	12.7	21.8	−22.0
20.95	7.66	14.0	21.3	−18.6
21.00	7.75	15.5	20.6	−16.3
21.05	7.82	17.1	20.0	−13.9
21.10	7.90	19.1	19.3	−10.4
21.15	7.98	21.6	18.5	−8.0
21.20	8.05	25.0	17.7	−5.4
21.25	8.13	30.6	16.9	−1.8
21.30	8.21	38.3	16.1	0.9
21.35	8.28	30.5	15.2	3.6
21.40	8.37	24.9	14.4	7.5
21.45	8.45	21.5	13.5	10.4
21.50	8.53	18.9	12.7	13.3
21.55	8.61	16.9	11.9	17.4
21.60	8.69	15.2	11.2	20.5
21.65	8.76	13.8	10.4	23.6
21.70	8.83	12.4	9.8	27.9
21.75	8.88	11.3	9.1	31.1
21.80	8.92	10.2	8.5	34.5

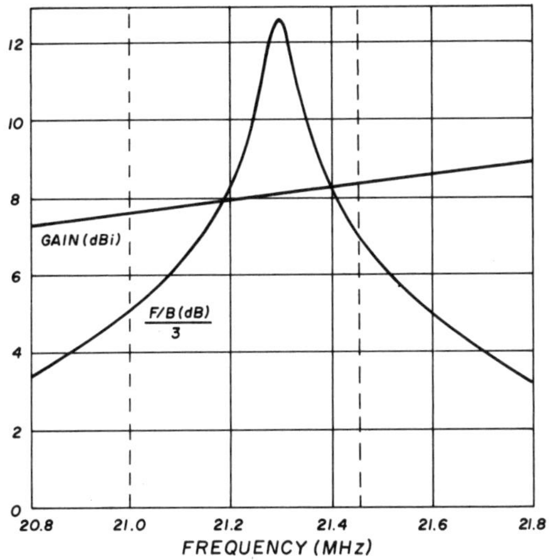

Fig. 8.15 — Three-element 21 MHz beam free space performance.

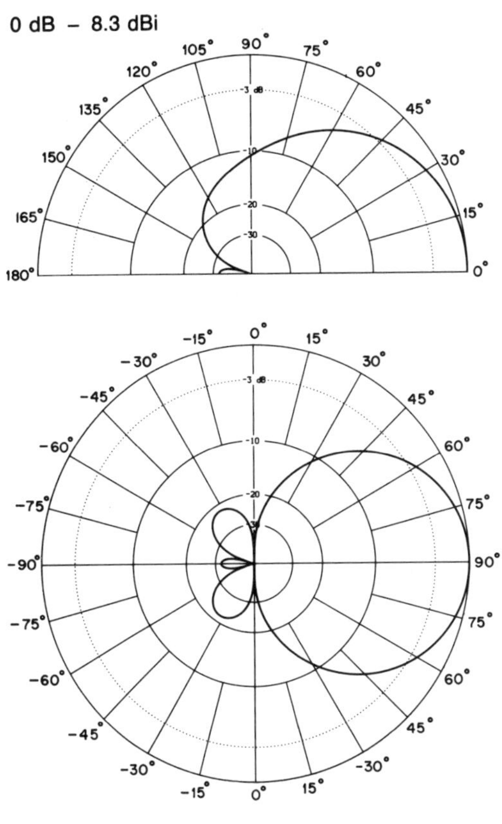

Fig. 8.16 — Three-element beam *H*-plane and *E*-plane patterns at 21.3 MHz.

Design and Performance Information for a Six-Element 21 MHz Yagi

Table 8.20 — Element positions and lengths for a six-element 21 MHz Yagi. Element diameters are 7/8 inches (radius = 0.00078899 wavelengths).

	position		l_{eq}		l
element	*(inches)*	*(λ)*	*(inches)*	*(λ)*	*(inches)*
reflector	0.0	0.0	273.41	0.49307	278.82
driver	85.2	0.15365	265.23	0.47832	270.32
D1	170.4	0.30730	245.55	0.44282	249.88
D2	255.6	0.46095	245.55	0.44282	249.88
D3	340.8	0.61460	245.55	0.44282	249.88
D4	426.0	0.76825	245.55	0.44282	249.88

Table 8.21 — Performance of a six-element 21 MHz Yagi.

freq. (MHz)	*gain (dBi)*	*F/B (dB)*	*resistance (ohms)*	*reactance (ohms)*
20.80	9.78	10.9	23.6	−31.9
20.85	9.92	11.9	23.4	−29.1
20.90	10.05	13.1	23.2	−26.4
20.95	10.17	14.3	23.0	−22.5
21.00	10.27	15.7	22.7	−19.8
21.05	10.37	17.3	22.5	−16.9
21.10	10.47	19.2	22.2	−13.0
21.15	10.55	21.5	22.0	−10.1
21.20	10.63	24.3	21.7	−7.2
21.25	10.71	27.7	21.5	−3.1
21.30	10.77	29.8	21.2	−0.1
21.35	10.83	27.8	21.0	3.0
21.40	10.89	24.4	20.9	7.3
21.45	10.93	21.7	20.8	10.5
21.50	10.97	19.5	20.7	13.8
21.55	10.99	17.8	20.8	18.3
21.60	11.01	16.3	20.9	21.7
21.65	11.02	15.0	21.1	25.2
21.70	11.01	13.9	21.5	29.9
21.75	10.99	13.0	21.9	33.6
21.80	10.95	12.1	22.6	37.3

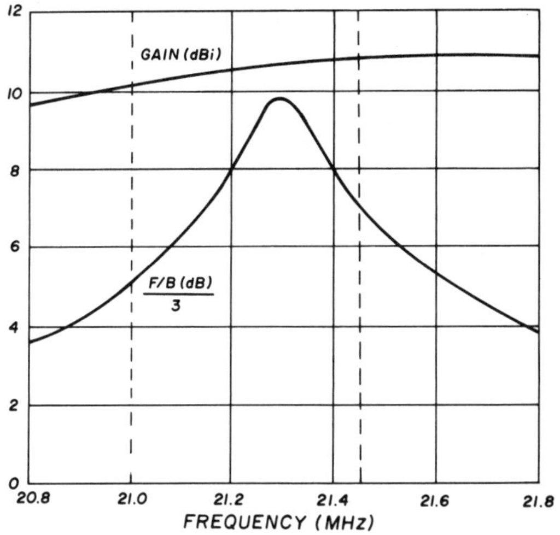

Fig. 8.18 — Six-element beam *H*-plane and *E*-plane patterns at 21.3 MHz.

0 dB – 10.8 dBi

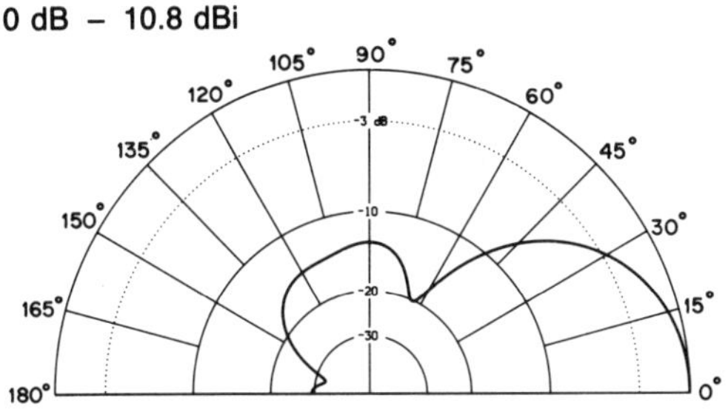

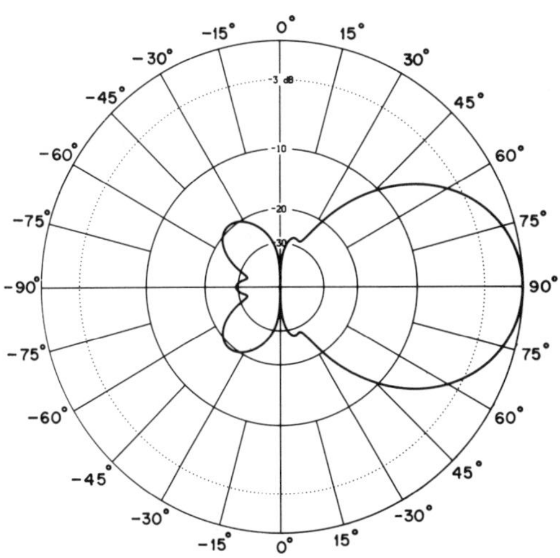

Fig. 8.17 — Six-element 21 MHz beam free space performance.

Design and Performance Information for a Three-Element 28 MHz Yagi

Table 8.22 — Taper schedule for 28 MHz elements.

section	length (inches)	diameter (inches)
00	1.94	1.902
0	1.0	1.000
1	69.0	0.875
2	variable	0.750

Table 8.23 — Element positions and lengths for a three-element 28 MHz Yagi. Element diameters are 7/8 inches (radius = 0.0010557 wavelengths).

element	position (inches)	position (λ)	l_{eq} (inches)	l_{eq} (λ)	l (inches)
reflector	0	0.0	205.27	0.49531	207.64
driver	57	0.13754	201.40	0.48598	203.69
director	114	0.27508	191.70	0.46257	193.79

Table 8.24 — Performance of a three-element 28 MHz Yagi.

freq. (MHz)	gain (dBi)	F/B (dB)	resistance (ohms)	reactance (ohms)
27.80	7.33	10.5	23.0	−27.3
27.90	7.46	12.1	22.4	−23.2
28.00	7.57	13.8	21.6	−20.1
28.10	7.67	15.9	20.6	−15.9
28.20	7.77	18.4	19.5	−12.6
28.30	7.86	21.8	18.4	−8.2
28.40	7.96	26.9	17.1	−4.6
28.50	8.06	33.7	15.8	0.1
28.60	8.17	27.2	14.5	4.9
28.70	8.27	21.8	13.3	9.0
28.80	8.39	18.3	12.0	14.2
28.90	8.50	15.7	10.9	18.5
29.00	8.61	13.5	9.8	23.9
29.10	8.70	11.7	8.8	28.5
29.20	8.77	10.1	7.9	34.2
29.30	8.79	8.6	7.2	39.0
29.40	8.76	7.2	6.6	44.8
29.50	8.65	6.0	6.1	49.8
29.60	8.44	4.8	5.8	55.8
29.70	8.11	3.6	5.6	60.8
29.80	7.66	2.6	5.6	66.8
29.90	7.08	1.5	5.8	71.9

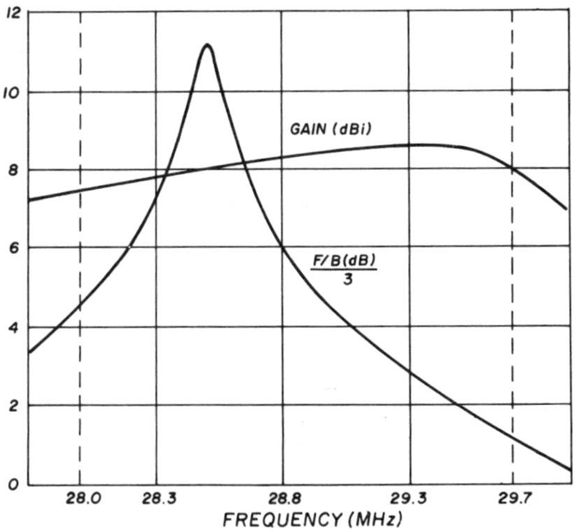

Fig. 8.19 — Three-element 28 MHz beam free space performance.

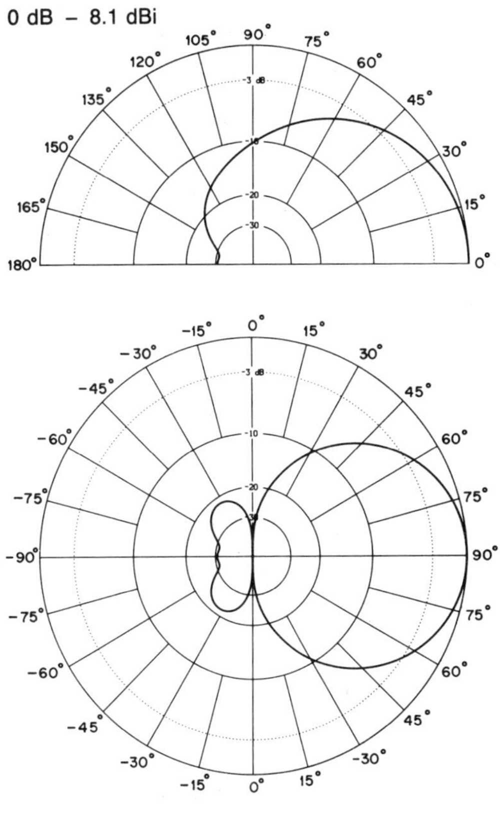

Fig. 8.20 — Three-element beam *H*-plane and *E*-plane patterns at 28.5 MHz.

Design and Performance Information for a Six-Element 28 MHz Yagi

Table 8.25 — Element positions and lengths for a six-element 28 MHz Yagi. Element diameters are 7/8 inches (radius = 0.0010557 wavelengths).

	position		l_{eq}		l
element	*(inches)*	*(λ)*	*(inches)*	*(λ)*	*(inches)*
reflector	0.0	0.0	204.20	0.49273	206.55
driver	63.6	0.15347	197.74	0.47715	199.96
D1	127.2	0.30693	182.27	0.43981	184.16
D2	190.8	0.46040	182.27	0.43981	184.16
D3	254.4	0.61387	182.27	0.43981	184.16
D4	318.0	0.76733	182.27	0.43981	184.16

Table 8.26 — Performance of a six-element 28 MHz Yagi.

freq. (MHz)	*gain (dBi)*	*F/B (dB)*	*resistance (ohms)*	*reactance (ohms)*
27.80	9.78	10.9	23.6	−32.7
27.90	9.98	12.4	23.3	−27.8
28.00	10.15	14.1	23.0	−23.8
28.10	10.30	16.2	22.6	−18.9
28.20	10.44	18.6	22.2	−14.8
28.30	10.56	21.9	21.9	−9.7
28.40	10.67	26.3	21.5	−5.5
28.50	10.77	30.1	21.2	−0.2
28.60	10.85	26.3	20.9	5.3
28.70	10.92	22.1	20.7	9.9
28.80	10.97	19.0	20.7	15.6
28.90	11.01	16.7	20.8	20.4
29.00	11.02	14.9	21.1	26.4
29.10	11.00	13.4	21.7	31.6
29.20	10.96	12.2	22.6	37.9
29.30	10.88	11.1	23.8	43.4
29.40	10.79	10.2	25.6	49.9
29.50	10.66	9.4	28.1	55.6
29.60	10.52	8.7	31.6	62.3
29.70	10.36	8.2	36.2	67.8
29.80	10.20	7.8	42.8	73.7
29.90	10.03	7.5	51.6	77.0

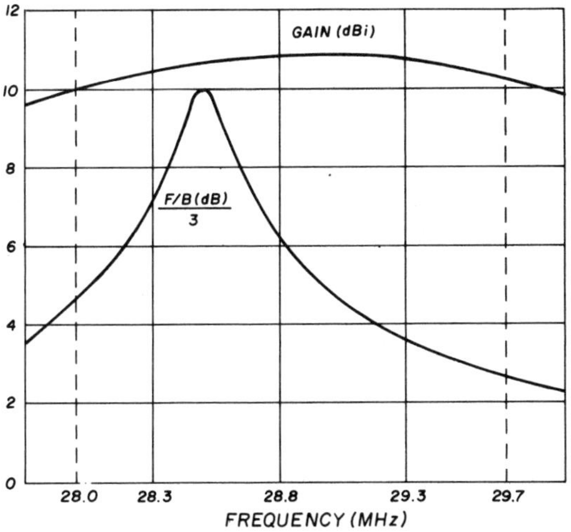

Fig. 8.21 — Six-element 28 MHz beam free space performance.

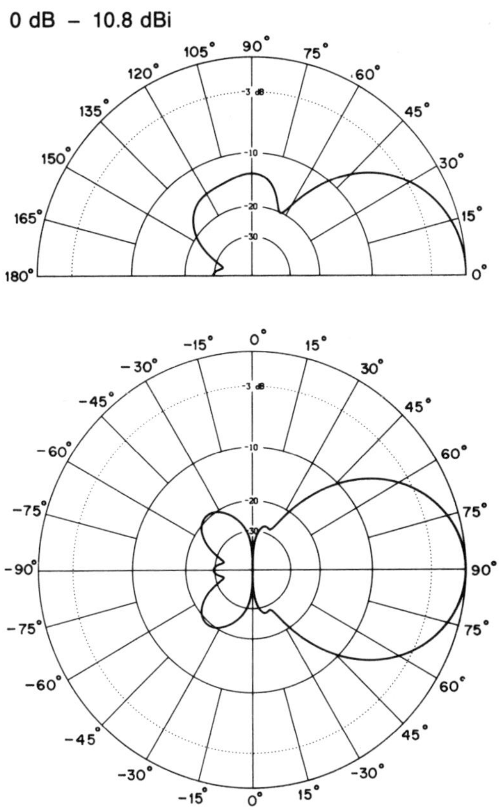

Fig. 8.22 — Six-element beam *H*-plane and *E*-plane patterns at 28.5 MHz.

Design and Performance Information for an Eight-Element 28 MHz Yagi

Table 8.27 — Element positions and lengths for an eight-element 28 MHz Yagi. Element diameters are 7/8 inches (radius = 0.0010557 wavelengths).

	position		l_{eq}		l
element	*(inches)*	*(λ)*	*(inches)*	*(λ)*	*(inches)*
reflector	0.0	0.0	205.14	0.49501	207.51
driver	76.3	0.18408	195.86	0.47261	198.04
D1	152.6	0.36816	175.70	0.42396	177.45
D2	228.9	0.55224	175.70	0.42396	177.45
D3	305.1	0.73631	175.70	0.42396	177.45
D4	381.4	0.92039	175.70	0.42396	177.45
D5	457.7	1.10447	175.70	0.42396	177.45
D6	534.0	1.28855	175.70	0.42396	177.45

Table 8.28 — Performance of an eight-element 28 MHz Yagi.

freq. (MHz)	*gain (dBi)*	*F/B (dB)*	*resistance (ohms)*	*reactance (ohms)*
27.80	11.31	13.0	34.4	−29.1
27.90	11.42	14.5	34.6	−24.6
28.00	11.52	16.1	34.7	−21.2
28.10	11.61	18.0	34.7	−16.7
28.20	11.70	20.3	34.7	−13.1
28.30	11.77	23.3	34.6	−8.5
28.40	11.84	26.8	34.5	−4.7
28.50	11.90	29.1	34.4	0.1
28.60	11.96	26.7	34.4	5.1
28.70	12.01	23.3	34.4	9.2
28.80	12.04	20.5	34.6	14.4
28.90	12.07	18.4	34.9	18.8
29.00	12.08	16.6	35.5	24.2
29.10	12.07	15.2	36.2	28.8
29.20	12.05	14.0	37.3	34.4
29.30	12.01	13.0	38.7	39.0
29.40	11.95	12.1	40.6	44.6
29.50	11.87	11.3	42.8	49.0
29.60	11.78	10.7	45.7	54.2
29.70	11.67	10.1	49.2	58.1
29.80	11.56	9.7	53.4	62.2
29.90	11.43	9.3	58.1	64.5

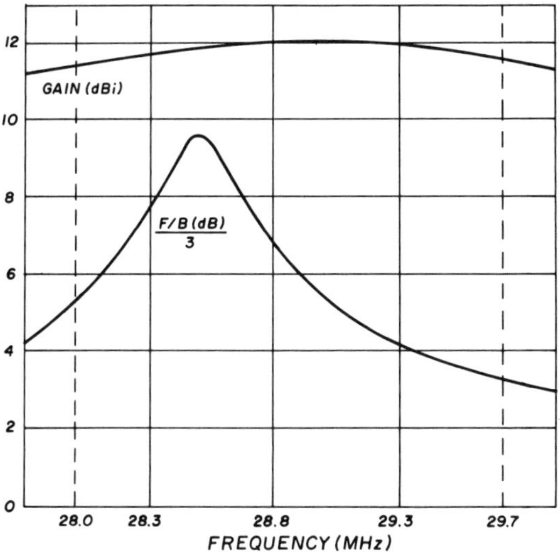

Fig. 8.23 — Eight-element 28 MHz beam free space performance.

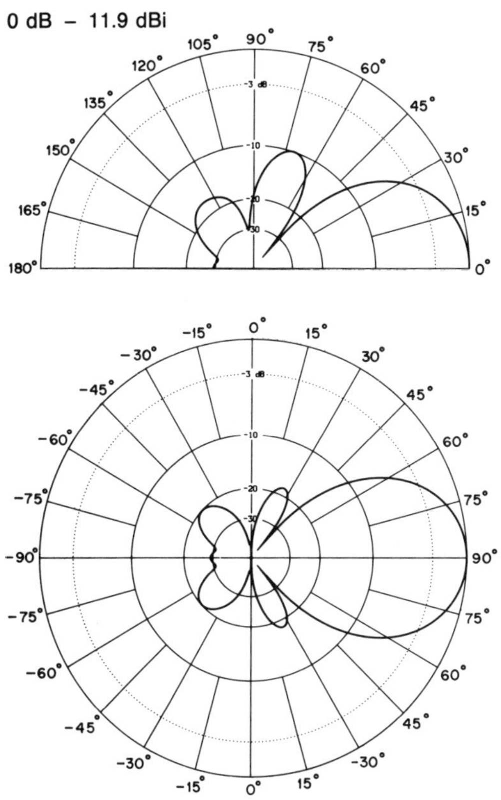

Fig. 8.24 — Eight-element beam *H*-plane and *E*-plane patterns at 28.5 MHz.

YAGI ANTENNA DESIGN

PROOF OF PURCHASE

FEED BACK

Please use this form to give us your comments on this book. Tell us what you liked best about the book, and what improvements you would like to see us make in future editions.

Name

________________________________ Call sign ____________

Daytime Phone (   ) ____________________ Age _________

Address ____

City, State, Zip ______________________________________

How long have you been licensed? _________________________

From ______________________

____

____

Please affix postage. Post Office will not deliver without sufficient postage.

Editor, Yagi Antenna Design,
First Edition
American Radio Relay League
225 Main Street
Newington, CT USA 06111

please fold and tape